T0385602

Oceans under Glass

Oceans in Depth

A SERIES EDITED BY KATHARINE ANDERSON
AND HELEN M. ROZWADOWSKI

Oceans under Glass

Tank Craft and the Sciences of the Sea

Samantha Muka

The University of Chicago Press CHICAGO AND LONDON

The University of Chicago Press, Chicago 60637
The University of Chicago Press, Ltd., London

Published 2023
Printed in the United States of America

32 31 30 29 28 27 26 25 24 23   1 2 3 4 5

ISBN-13: 978-0-226-82413-0 (cloth)
ISBN-13: 978-0-226-82414-7 (e-book)
DOI: https://doi.org/10.7208/chicago/9780226824147.001.0001

Publication of this book has been aided by a grant from the Bevington Fund.

Library of Congress Cataloging-in-Publication Data

Names: Muka, Samantha, author.
Title: Oceans under glass : tank craft and the sciences of the sea / Samantha Muka.
Other titles: Oceans in depth.
Description: Chicago : University of Chicago Press, 2023. | Series: Oceans in depth | Includes bibliographical references and index.
Identifiers: LCCN 2022020602 | ISBN 9780226824130 (cloth) | ISBN 9780226824147 (ebook)
Subjects: LCSH: Marine aquariums. | Coral declines. | BISAC: PETS / Fish & Aquariums | NATURE / Environmental Conservation & Protection
Classification: LCC SF457.1 .M85 2023 | DDC 597.177073—dc23/eng/20220617
LC record available at https://lccn.loc.gov/2022020602

♾ This paper meets the requirements of ANSI/NISO Z39.48-1992 (Permanence of Paper).

Contents

Series Foreword

Oceans in Depth

Samantha Muka's *Oceans under Glass* launches a new series by the University of Chicago Press, Oceans in Depth. The series publishes works that put the ocean at the center of our narratives about the past. When we move beyond narrow coastal slices to consider the oceans in their depths, we gain new dimensions to our histories, both in the modern era and through deep time. To build these fuller accounts, the series adopts a broad definition of historical writing. Contributions to this series emerge from a variety of disciplines and perspectives, such as history of science or technology, historical geography, anthropology, environmental history, art history, literary history, and nature writing. While anchored in rigorous scholarship, the works speak to broader academic, student, and general audiences.

The ocean has profoundly shaped human existence as a space of sustenance, industry, and exchange as well as a source of knowledge, myth, and imagination. The complex interactions between humans and the ocean, though ancient in origin, have tightened over time and multiplied with globalization. The importance of these interactions today—in terms of climate, health, economy, food supply, recreation, coastal habitation, and many other areas—prompts new and urgent attention to understanding our past relationships with the ocean.

How does a book about aquariums constitute ocean history? In the last century, the aquarium as a spectacle—ever larger, ever more immersive—has brought the ocean environment to the view of millions, sustaining public imagination about an underwater world. Samantha Muka's story of "tank craft," the technical practices that developed to maintain these artificial environments, reveals their complex relationships to our understanding of the ocean itself. Ranging from underwater photographic techniques to the scaled-up simulation of Florida's Discovery Cove, Muka's stories go beyond the technology of the aquarium to uncover the communities that created it: hobbyists, commercial exotic fish breeders, public aquarium staff, laboratory technicians, and others whose mostly informal communications amassed knowledge and skill. These networks are surprisingly varied and international and reach far inland, a picture of the ocean's presence across many cultures. The tank craft tradition and its many practitioners tell us something else unexpected. Although the secrets of marine life usually summon to mind scientists on ships hauling specimens from the deep, or marine biologists peering at specimens in the laboratory, the idea of modeling the ocean developed in critical ways beyond the realm of academic specialists. Experts and enthusiasts created tanks to imitate the open ocean currents that sustain jellyfish, and a series of tanks to support coral reef species through their complex life cycles. Important knowledge of the volumetric ocean—that is, the ocean's changing layers below its surface—emerged from efforts to keep diverse creatures alive and reproducing under glass.

Oceans under Glass shows how to explore new dimensions of our ideas of the ocean and its many places in our lives. Employing a combination of archival research and ethnography, Muka's account is a guide to the hidden worlds of tank craft. The consequences of these stories are profound. We know oceans through work, experiment, and the consumption of spectacle. Yet at the same time, the mimicry of the ocean confronts us with the hubris of human efforts to contain the natural world. Now and into the future, the ways we approach the oceanic environment may depend on lessons acquired tinkering with tanks.

Katharine Anderson
Helen M. Rozwadowski

Preface

I keep coming back to a single picture.

The ocean is a vast expanse. Most estimates show that humans know very little about what is in the ocean. The National Oceanic and Atmospheric Administration (NOAA) estimates that we've mapped, observed, and explored less than 20 percent of the ocean.[1] The terms here aren't necessarily overlapping; we could have observed a single space for a very short period of time and still consider it part of that 20 percent. But if we use a narrow definition of exploration to mean time spent examining and developing intimate knowledge of the ecosystems that make up the ocean and how they function over extended periods, then you might say that we know very little about the marine environment. Estimates put the number in a startling low range. This isn't unknown or unacknowledged by marine researchers. In a bid to build international networks of exploration and to better understand not just what a place looks like but to get a sense of the biological diversity of the ocean, over seventy countries and one thousand scientists participated in the Census of Marine Life (CoML) between 2000 and 2010.

The CoML was meant to survey as many of the ocean's inhabitants as possible. The project linked far-flung researchers into a network of like-minded explorers. Over a decade, that network explored and shared data about marine environments, many that had been under- or unexamined before that time. The CoML reported findings on new species and the first explorations and observations of whole ecosystems. The development of a Global Marine Life Database collated the found organisms into a database, and the first images of those new species became available online, not just in research papers. The CoML expanded knowledge of what was in the ocean, but it also allowed new visualizations through the digitized species and field images provided to the public.[2] Vibrant images of sea stars and jellyfish can be accessed at the CoML website. And mixed

FIGURE 0.1 Researcher Neil Bruce of the Museum of Tropical Queensland studies specimens in lighted aquarium on Lizard Island Reef. Photo: Gary Cranitch, courtesy of Queensland Museum.

among them are field images. It's one of those images, featuring a field site in Queensland near sunset, that I can't get out of my mind.

In the image, Neil Bruce is standing knee-deep in water on Lizard Island Reef peering into a lighted, floating aquarium. The CoML researcher is intent on his observations and appears completely alone in the expansive ocean. Beyond the beauty of the image, what is striking is the framing. The light coming out of the aquarium is the main source of illumination, and it expands not just toward the camera but outward underneath the water. While it is beautiful, it is also an illustration of the integral role of aquariums in facilitating the gathering of marine knowledge. The aquarium is essential in the study of the marine world, making it possible for a researcher to view a slice of the environment in a simple, square space.

This is the only time that an aquarium shows up in the images from the CoML. Of the thousands of images released online from the project, many of which were captured using aquariums, this is the only one that contains that integral piece of technology that makes much of the work possible. Aquariums, their maintenance, and the work done with them are nearly invisible, even when in plain sight. This book is about the development of specialized aquariums, how they became integral to the study of the ocean, and how the reliance on that technology tells us much about our evolving relationship with the marine world.

Aquariums are three-dimensional, primarily clear structures that allow the observation of or experimentation on captive aquatic organisms. Humans have been building enclosures for aquatic organisms, for both aesthetic and aquacultural purposes, for millennia. Most of these pools consisted of rock enclosures permeable by nearby water sources that trapped growing fish and made harvest possible. The permeability of those spaces made it possible to get an influx of natural food and younger organisms for continuous restocking; they required very little attention by those using them.[3] But the modern aquarium is something different. The Industrial Revolution resulted in the development of modern glass-making techniques and a subsequent drop in glass prices in the mid-nineteenth century.[4] This prompted those interested in keeping aquatic organisms in captivity to build the earliest aquatic closures meant to be viewed from both the sides and above. The design became almost instantly popular with marine researchers, hobbyists, and those hoping to show these spectacles to the public. The rise of the laboratory, home, and public aquarium was almost simultaneous.

As each of these communities developed aquariums to suit their purposes, they exchanged techniques and technologies that proved useful throughout the aquarium-keeping network. The simplest aquarium contains water, a source of oxygen, and an inhabitant. Think about the round glass tank usually holding a goldfish or betta fish won at a fair. If you've ever taken one of those plastic-bag-bound fish home, you know that the simplest setup is often far from easy to maintain. An individual fish still requires an environment calibrated for balanced light, food, and water chemistry. The uninitiated usually end up flushing their mistakes and desperately steering their children away from that booth at the fair the next fall. But there exists a community of aquarium users who take up the challenge of creating these environments and develop intricate tank systems to perform a variety of tasks to keep an increasingly diverse group of organisms and ecosystems in captivity. Hobbyists, public aquarists, and research scientists all work to develop these systems, and through the exchange of ideas throughout these communities, they can now maintain a wide array of marine organisms indefinitely.

Through the development of these tank systems, humans have come to know the marine world they seek to replicate in miniature. The process of simulating an environment, of trying to understand what it should contain for success in keeping captive organisms, sets up a feedback loop of knowledge about the sea. Developing a specialty tank requires some knowledge of the larger marine environment. This can be something as simple as the salinity of the water or the location from which you collected an organism. If it came from a reef in the Caribbean, aquarists have some general information about the water temperature and chemistry of that location. They start by recreating that chemistry and temperature, as close as possible, in the aquarium. But the marine world is visually inaccessible for long periods to humans, and aquarists guess about certain variables such as age of sexual maturity and natural history. What does this organism eat? What is normal feeding behavior? Mating behavior? Are the color changes seen in the aquarium "normal" or caused by the stress of captivity? The longer the organism is in an artificial environment, the more questions are answered, and also asked, about that animal in its native habitat. Captive specimens both confirm field observations and raise new questions to be answered by that research.

The study of the aquarium also highlights the importance of nonstandardizable technologies and tinkering to modern biological research. Many tools for studying the natural world become standardized over time—they can be calibrated by large groups so that everyone is sure to be measuring using the same starting point. Over time, those tools become "black boxed," meaning the user need not understand the mechanisms by which the tool works to achieve acceptable output.[5] The aquarium is not such a scientific tool. Building tanks for two genetically identical organisms in two separate locations might result in two very different systems. Each aquarist tinkers with the system based on both the internal (temperature, chemistry, needs of inhabitants) and external (light, season, temperature, sound, etc.) conditions of the tank. The confluence of internal and external variables makes it difficult to standardize tank systems. This inability to standardize and streamline the process of tank building has resulted in a lack of patents for specialized tanks and a reliance on basic techniques without formalized and foolproof instructions. While it is possible to develop general parameters, most builders realize that tinkering is integral to the process of developing the systems and that protecting a patent would be nearly impossible.

Focusing on the aquarium as a tool for understanding the submarine world can tell us not only how we know the ocean but who contributes to that knowledge. While the aquarium is integral to academic biological research, it is not a scientific tool isolated in the laboratory and used exclusively by academic biologists. Instead, the development and deployment of systems to mimic and study the marine environment occurs in homes, public entertainment facilities, fisheries laboratories, commercial spaces, and academic environments. This wide array of aquarium users contributes to marine-biological knowledge production through the exchange of information about the optimal operating conditions for various tank systems and the life histories of their tank inhabitants. Tracing the exchange of information shows that this reliance on a nonstandardized technology means that marine biology as a field has permeable boundaries. The extensive network of nonacademic actors consists not of amateur scientists or "citizen scientists" contributing data points to a larger academic endeavor but instead of a range of experts contributing to the advancement of more basic knowledge about the sea. By tracing that exchange of information, we can see how important this wide community of aquarium users is to the development of marine-biological knowledge.

I keep coming back to Bruce's concentration, not on the sea around him, but on the interior of the lighted aquarium. Too often, aquarium work is not mentioned when researchers talk about what we know about the sea. Press releases and methods sections skim over tank details, giving the impression that all work on marine organisms is done in open water. But this book will show that the aquarium is integral to our understandings of the sea and that our knowledge of what constitutes the open ocean is often predicated on the specialized tanks that model it.

This study contributes to a growing body of literature that sits at the intersection of environmental and biological history and science and technology studies, often referred to as *envirotech*.[6] This field uses traditional science and technology theory to analyze the history of human interactions with and constructions of the environment. Scholars have noted that the engineering spirit of modern biological and ecological research, through the development and use of technological innovation, blurs the lines between knowing and making.[7] The aquarium is representative of this blurring of boundaries. Tank parameters shift as aquarists learn what does and does not work for maintaining captive organisms. That shifting represents knowledge production about both the organism and their environmental interactions and needs. Studying the historical development of aquariums can tell us much about who can contribute knowledge through this type of engineering and how working with these captive systems has contributed to and shaped knowledge about the marine world.

This book contributes to envirotech discussions in two important ways. The first is through interdisciplinary methodologies. The development of specialized aquarium technology is nonlinear and dispersed among a wide network of users. Unlike other formal experimental technologies that historians of science have traced, such as model organisms or electron microscopes, aquarium development doesn't have a central location. In addition, the nature of the technological transfer of information is ephemeral: most knowledge exchange occurs through informal interactions. This inability to anchor the technology in a single institution, or the knowledge exchange in a specific academic journal or conference, makes it particularly difficult to rely exclusively on traditional historical methodologies, such as oral histories and archival sources. While I use these traditional sources when possible, I also use observation, interviews, and other sociological and anthropological methodologies to trace the development and growth of aquarium technology. This mixed methodology has resulted in a history of biology that is not bound to traditional academic spaces or publications.

In addition, this work introduces historians of biology to a wide community of relatively unknown actors. Readers will recognize some of the traditional figures in the history of biology—don't worry, Darwin and Haeckel get a mention—but the majority of the chapters introduce a vast array of men and women who have never had their history told but who are vital to the development of marine-biological knowledge. There are two critical reasons that focusing on aquariums brings these actors into focus. The first is that the history of biology has traditionally had a terrestrial focus. Histories of organisms, environments, and institutions commonly focus on land-based biological questions. The history of marine biology, a small and scattered field, is becoming increasingly popular, and this work contributes to that growing body of literature.[8] The second reason is that my focus is not exclusively on academic biology, but instead on the wider network of aquarium users including hobbyists, public aquarists, and fisheries biologists. In my previous research, I have sought to find this transdisciplinary network at institutions such as public aquariums or marine stations. And they do sometimes show up in those spaces. But the true picture of this network is only revealed if one follows the aquarium technology and leaves the traditional understandings of institutional or disciplinary boundaries aside. To see the vast network of actors involved in marine-biological knowledge production, it is important to follow the technology.

This book's chapters trace the historical arc of individual tank systems. Taken together they tell us much about the way that knowledge is created and travels through a wide network of users. Chapter 1 provides context, both theoretical and historical, for the study of tank technology. I look at the way my study fits into or challenges current thinking on tinkering, scientific networks, and environmental visualization. The aquarium is a hard technology to trace, and this chapter gives the larger history of the main groups who use the aquarium, why they work with them, and how I follow craft and tacit knowledge. This chapter provides context for the academic arguments in which I am engaged, but it should not be too dense for those interested in tanks who do not have a historical or theoretical background.

Chapter 2 looks at the development of the photographic tank. Today, digital photography allows tank users to share images of their systems quickly and efficiently. If you want to show off a new system, you upload a photo online for likes and comments. If you want help diagnosing a problem in your tank, one of the first questions is going to be "do you have a photo?" Aquarium photography has come a long way since the earliest attempts to share images taken through plate glass. At the turn of the twentieth century, photographers were doubtful that underwater photography was even possible. The development of specialized equipment to capture images in aquariums came from the intersection of two hobbyist communities—photographers and aquarium keepers. The spread of this tank technology into public aquariums and academic research shows the importance of aquariums as visual objects. Aquariums allow users to build imagined ecosystems, and the photographic tank disseminates those images throughout the world. Photographic aquariums are temporary tools for capturing images, but other tanks are constructed as essential for the perpetuation of organisms in captivity.

Chapter 3 focuses on one of the first systems for maintaining highly specialized organisms: the kreisel tank. Jellyfish and other gelatinous zooplankton live in the open ocean in the water column. This environment is very unlike the average glass tank: it has gentle, undulating water movement that slowly moves jellyfish without agitating their delicate structures. Early physiologists working with jellyfish to explore nerve physiology constructed tanks to maintain these organisms for extended periods in hopes of understanding nerve regeneration and growth. Public aquarists refined the tank structure and feeding schedule of a wide variety of species, exponentially increasing the number able to be displayed in captivity. Chapter 3 focuses on the gendered division of labor in tank craft. Historically in kreisel work, advancement in tank design was done by men and advancement in technique by women. The division of labor required to maintain jellies in captivity, and the gendered aspect of that division, should be understood to capture fully the groups of people contributing to tank craft. The kreisel system is a tank that seeks to mimic the open ocean current, and it results in a space that is relatively empty except for the single species contained therein. But other tanks contain entire ecosystems.

In chapter 4 I turn to the development of the reef tank. Reef tanks are tank systems meant to be a functioning reef ecosystem, complete with various species of coral, invertebrate, and vertebrate species. The earliest tanks dedicated to reef ecosystems were developed by hobbyists in the Indo-Pacific, but the craze quickly spread to hobbyist groups and public aquariums throughout the world. Today, the knowledge gained from keeping coral in captivity is being used to breed assurance populations for reef rehabilitation to try to help reefs suffering from the effects of climate change. The history of reef tanks shows us how aquariums serve as experimental systems that force the user to think about species interaction in both the wild and captivity. In addition, the development of these tanks demonstrates the ways that expertise is negotiated between various groups in the network of users.

Finally, chapter 5 examines the newest push to develop specialized tanks: marine ornamental breeding systems. Today, the value of marine ornamental species is on the rise. The current climate crisis, coupled with centuries of unwise fishing methods, has greatly depleted the number of reef fishes available both to maintain healthy reef ecosystems and to supply the aquarium industry. This crisis is leading to the development of commercial and government endeavors to breed the most popular ornamental species. Beginning with the successful breeding of the clownfish in the early 1970s, there has been a concentrated effort to produce captive-bred fish for the aquarium market. Recently, a network of researchers working in academic fisheries biology, public aquariums, and commercial fish farms has successfully closed the cycle on some of the most-sought-after fishes in the field, including yellow and blue tangs. This chapter looks at the history of ornamental fish breeding systems and highlights the current issues with the species being studied. Unlike the previous chapters, this chapter follows the process of developing systems in real time to examine how researchers navigate difficulties and share information throughout the network. By bringing the work up to the present, we can see how tank craft continues to bind together marine researchers from a wide variety of fields.

Each chapter has its own lessons, but taken together they point toward an important aspect of marine science: the acceptance of a variety of forms of expertise, and the ability to share knowledge through a simple technology, continues to advance the study of the marine world. In the current climate crisis, the threat to marine biodiversity is enormous. The question is often posed: How can we stop the degradation of the environment and engage a wide community of people in the restoration of the seas? But this question does not account for the ways that a wide community has been working to understand the ocean more fully. This network of marine researchers and tinkerers has spent the last century figuring out how, as one reef tank builder says, to "open our box up and let the open ocean in."[9] The creation of knowledge about the ocean relies on these boxes (aquariums); they allow users like Dr. Bruce to place oceans under glass to understand them more fully. To save oceans, we must fully understand who creates knowledge about them and how they gained that knowledge.

1

Aquarium Craft
Replicating Oceans under Glass

Mary Akers. You probably don't recognize the name, but her story is amazing. Akers is a crab enthusiast and one of only five people worldwide who have managed to breed land hermit crabs, *megalopa*, transitioning them from eggs into water-dwelling juveniles and back onto land to live as mature hermit crabs. This is the type of accomplishment that I mention at parties and wait for people to get excited. *She closed the cycle on hermit crabs! Cool, huh?* The blank stares don't deter my excitement and usually lead me to offer more information. Akers isn't an academic biologist. In fact, it is reported that she struggled in freshman biology, decided that science wasn't for her, and became a pottery artist. She returned to breeding crabs out of curiosity and "obsession" after her children left for college. She has an Etsy store where she sells "products for the well-fed, fashionable hermit crab" including homemade ceramic pool ramps, food dishes, and dried mealworms and other foodstuffs for hermit crabs.[1] *Mary Akers is so awesome. Right?!*

Akers blogged extensively while developing her methods with *megalopa* and continues to do so as she works to breed other species in captivity. Her blog *In the Crabitat* outlines tinkering with the habitat in which the crabs live, feeding and care techniques, details about the behavior of her crabs, and her thoughts on the ethics of captive breeding. She includes images of her tanks, her crabs, and details of her feeding regimen. She states that "A big part of why I'm doing this is to make it easier for those that try after me—creating a record of what worked and what didn't to take away some of thc guesswork for others—but day-to-day that goal gets subsumed as I fret and worry and try something new when the old stops working."[2] According to Akers, she bases her tinkering on knowledge developed throughout her lifetime. On September 1, 2017, she wrote,

> Anyway, a lot of what I'm trying is based on what I know about the ocean, how nutrients travel through the water column in day and night

> cycles, and so far so good. In 1998 I co-founded a marine ecology study abroad program in the Caribbean (Dominica) and in the early 90s I lived and worked at a marine ecology school in the Turks and Caicos Islands which had tons of hermit crabs and I observed them on a day-to-day basis in their natural habitat, in addition to diving in Caribbean waters and studying ocean ecosystems. I like to think that because I have a base of somewhat-relevant knowledge, I can help advance the process of captive breeding for the good of hermit crabs all over the world.[3]

Eighteen months later, her work paid off.

In November 2018, Akers announced on her blog that she had successfully transitioned captive-bred hermit crabs to land (closing the cycle from fertilization to mature adult). She spent a lot of the month preparing her sandy beach to be overrun by naked baby crabs looking for tiny shells to wear and trying to find those little crabs as they made their transition from water to land. On a particularly hard day, she used a mantra to get through the maintenance required to keep hundreds of tiny hermit crabs alive:

> "You aren't playing God, you're playing Ocean." Adopting that slant helped some, especially when I thought about all the times I've been in or on the ocean. Sometimes the vast ocean is comforting, a cradling, crackling, salty womb, but other days it's wild and dirty and scary, completely indifferent to my piddly little life to the point of seeming mean. On those days when I'm struggling hard and losing some of the zoeae without knowing why, or fumbling in my efforts and causing unintended issues, I just tell myself, "You're the ocean, Mary. And today the ocean is dirty and mean."[4]

In Mary's mind, the tank system had become the ocean and Mary herself a force of nature.

My enthusiasm for Mary's story is centered not on the quirkiness of it but instead on the fact that her success through tinkering and the dissemination of her resulting techniques is typical of marine knowledge developed with tanks. Akers's work with *megalopa* is a good example of the way that small-scale, niche hobbyist experimentation with an aquarium tank can affect larger discussions and understandings of the marine world. Akers is a hobbyist who has developed a method to do something many academic and public aquarists have not. But she is not isolated. Instead, her blog gives a clear accounting of her tinkering with both her tank system and the inhabitants of that system. And her work is known and spread throughout a larger marine network. She states on her blog that readers have sent her food or other care items for her crabs and her products are well liked on Etsy. On March 11, 2019, one customer remarked, "Wonderful little bowls! Both beautiful and useful. Great for testing new individual foods and gauging how popular they are with the crabs. Thank you!"[5] I was informed by a friend that the only reason her classroom hermit crabs have survived is because Mary is part of an online Facebook group dedicated to hermit crab care. Mary is also known in academic communities. In a feature article about her work, academic biologists interviewed said that they knew about Akers's work and found it "impressive." Akers's work and her connection to the larger marine network can tell us much about the importance of tanks to knowledge about the sea.[6]

Aquariums offer an opportunity to study marine organisms and the ecosystems in which they evolved. To build a captive ecosystem, users must pay attention to everything that is known about the natural world. When a specimen is initially collected, the information used to set up tank parameters—including water temperature, salinity, lighting, and food—comes from what we know about the location from which the specimen was taken. The tank is set up according to this knowledge, but there are always unknowns regarding the life cycle and ecological requirements of organisms. Through the process of trying to maintain these organisms, the aquarist discovers and refines those details about the interactions between the organism and its environment in the wild. Early zoological parks contributed to knowledge about those animals they sought to maintain. "The problem of keeping expensively purchased animals alive led zookeepers to pay increasing attention to animals' needs in the way of food and shelter, their breeding seasons and habits, and their requirements for space and activity—all issues that contributed to expanding natural history in the direction of functional knowledge of the living animal."[7] The knowledge gained from long-term captivity can inform those who hope to better understand the organisms in their native environment. For instance, changing the parameters of the tank can cause changes in the organism and reveal the catalyst for certain behaviors in the wild.[8] The aquarium and the ocean are linked through this feedback of knowledge, and we cannot know one without the other.

The aquarium is integral to the study of the ocean and its inhabitants, but it is problematic. Unlike other basic scientific tools, these systems cannot be standardized. An aquarium seems like a simple technology. It is a transparent enclosure containing water, a filtration system (even if this is manual) and an organism. But that simplicity belies the complex negotiations between a wealth of variables that must be balanced before the goal—survival and thriving of the inhabitants—can be achieved.

It can take years for the process of keeping a new organism in an aquarium to be worked out for a single tank. And to complicate this, the process used in one tank cannot necessarily be replicated in another. Aquariums require craft knowledge—a type of knowledge developed through hands-on tinkering and the combination of both scientific and anecdotal knowledge about the systems you hope to mimic. Just as Akers relies on her past of walking the beach to build tanks for hermit crabs, the craft of building and maintaining tanks combines a variety of ways of knowing to produce a useful model of the marine world.

The technology requires communication between a wide range of users. The aquarium is a technology that can be scaled up or down in both size and detail depending on the goals of the user. But that scaling is not dependent on access to professional or high-end technologies. In fact, aquariums are one of few technologies used in research laboratories that look similar and use the same components as those in public aquariums and homes. The ability to get the system working, to produce knowledge from replicated marine environments, can come from anyone in the marine network. This contributes to the pool of information needed for all of us to know the ocean. Producing knowledge about the marine world with aquariums doesn't just invite a conversation between disparate professional and social groups; it requires it. Akers belongs to a group called hobbyists, but there are professional academic biologists, public aquarists, fisheries biologists, and commercial aquarists in this network as well. Tanks and the craft of keeping them draw this large network together, and it is the exchange of tank craft information that produces knowledge about the wider marine world.

This network of aquariums, connected by aquarists sharing tank craft, places oceans under glass. Aquariums are the only way to study marine organisms and ecosystems over the long term. Humans have tried to build underwater areas for constant observation with varying success. Aquariums provide space for aquarists to interact with a wide range of organisms in captivity over long periods. Building these tanks requires tinkering, but it also requires the aquarist to have an idea about the nature of the wider ocean they are seeking to replicate. Akers speaks about her interactions with the ocean throughout her life and her perceptions of the ocean as sometimes gentle and fertile but sometimes angry and deadly. She imagines herself as a sort of Poseidon figure, presiding over the watery depths of her glass-enclosed ocean. This combination of natural history knowledge, interactions, and imagination helps aquarists build tanks, and it also affects the interpretation of their work and, eventually, the application of that work to the open ocean.

This book examines the history of oceans under glass by tracing the development of specialized tank design. By studying tanks, I hope to develop a better understanding of how craft knowledge develops and spreads through disparate communities and the effect that tank building has on the way that we come to know and conserve the ocean. By studying the aquarium, we can better understand how a ubiquitous but overlooked tool, and a wide group of users, shapes our relationship with the sea.

Crafting Oceans under Glass

Creating aquariums that function in the way the user desires requires trial and error. The basic aquarium is one that has glass walls, water, and a filtration system of some sort. In the simplest goldfish bowl, the filtration system is a manual water change every week or so. This is as intricate a system as some people care to maintain, and it does the job: your goldfish thrives. But the tank systems I discuss in this book have specialized purposes, and the requirements for a functioning tank system become increasingly difficult. Successfully getting clear images of animals, keeping jellyfish alive, keeping coral reefs alive, or getting recalcitrant fish to breed takes both in-depth knowledge of the organism and its natural life history and dedication to tinkering with the system until something works. Every chapter of this book highlights the role of tinkerers in the development and spread of aquarium technology and use. But what does it mean to tinker and why is it integral to aquarium work?

The process of tinkering involves the continued shuffling of available components until a system works in the desired way. Anthropologist and ethnologist Claude Levi-Strauss differentiates tinkerers, called *bricoleur*, from the craftsman, engineer, and the "odd job man" or handyman. According to Levi-Strauss, the *bricoleur*'s skills are not defined by a given project but by the ability to improvise with the same set of components repeatedly.

> The "bricoleur" is adept at performing a large number of diverse tasks; but, unlike the engineer, he does not subordinate each of them to the availability of raw materials and tools conceived and procured for the purpose of the project. His universe of instruments is closed and the rules of his game are always to make do with "whatever is at hand," that is to say with a set of tools and materials which is always finite and is also heterogeneous because what it contains bears no relation to the contingent result of all the occasions there have been to renew or enrich the stock or to maintain it with the remains of previous constructions of destructions.[9]

Levi-Strauss argues that the bricoleur requires some specialized knowledge and tools, but only to a point, because they are adept at applying what they already know to a given problem.

The differentiation of the tinkerer from the engineer or craftsman is not only due to the nature of the tools and resources used but also to the eventual outcome. In 1979, biologist François Jacob applied Levi-Strauss's concept of the tinkerer to the study of evolution. Jacob claimed evolution itself is a tinkerer—working with the building blocks of life but combining and revising them over eons with new, but not necessarily well-engineered, results. According to Jacob, not only are the limited tools of the tinkerer important but two tinkerers could solve a problem in drastically different ways. "Unlike engineers, tinkerers who tackle a problem are likely to end up with different solutions."[10] This means that tinkering is both problem-solving and a form of self-expression and creativity, and the results are as unique as the individuals who create them.

A problem solved through tinkering is one that demonstrates both the individual thought process of the tinkerer and the materials with which they worked. Historian of technology Kathleen Franz calls tinkering an "act of creativity" that relies on "spatial thinking and 'fingertip' knowledge rather than formal education or training in science and engineering."[11] This creative side of tinkering can be seen in sociologist Caitlin Wylie's work on technicians in paleontological laboratories. Technicians are often volunteers using found objects and inventing techniques to remove surrounding stone and to clean and shape fossils. According to Wylie, the laboratory contains

> many scattered paper cups filled with dental scrapers, paintbrushes, plaster, glue, broken fossils, or this morning's coffee. Pneumatic engravers, delicate handheld jackhammers designed for writing on metal, are plugged into air-lines at every workstation. Fossils wait for attention in various cardboard boxes (some emblazoned with food labels or shoe brands) or cushioned on sandbags made of burlap, old jeans, or recycled socks. Sandblasting equipment, often housed inside a homemade chamber of clear plastic with holes cut to admit the technician's hands, may occupy one table. These striking collections of objects—obtained from "dumpster diving" expeditions and dentists' donations as well as from science supply companies—give the impression of a place that is simultaneously a hardware store, a tinkerer's workshop, and an artist's studio.[12]

Many of these tinkering technicians claim their work as artistic and see their role as mainly creative. Wylie states that, "preparators' claims to art and creativity are best understood as justifications for their diverse and individualized problem-solving practices."[13] It is the tinkering nature of their endeavor that creates such diversity in techniques and outcomes, and this diversity is seen as artistic.

Scholars have studied two separate groups of tinkerers: consumers and laboratory technicians. The consumer tinkerer is a community that seeks to enhance or turn a preproduced technology into an object that functions in different ways. Franz's work revolves around the tinkering done by early consumers of automobiles in the United States. According to Franz, the earliest car buyers formed communities, not just as those individuals with access to a new technology but in a community of people who had to learn quickly how to shape and repair broken machines.

> Automobile ownership gave American consumers a part to play in the national dialogue on technology. Consumers did not simply drive the automobile, they repaired, tinkered with, and intervened in the design of the machine. Early automobile travel was difficult. Driving, and especially long-distance touring, required basic mechanical knowledge in order to run and repair the machine. The difficulties of early motor travel, when added to the middle-class expectations of economy and comfort, inspired travelers to learn something about mechanics and to tinker with the design of the car.[14]

Recently, social scientists and historians have highlighted the tinkering communities that develop around medical technologies. They show that technologies that interface with bodies often require calibration with the individual, something that individuals do to make their quality of life better. Other communities that tinker build or alter technologies that have been denied them. As the cost of diabetes technology and insulin rise, tinkerers in the type 1 diabetic community have built or altered technologies to provide affordable relief to those suffering from the illness.[15]

Tinkering is also seen as a form of gendered expression. While Franz has highlighted the way that women tinkered with cars to assert independence and the ability to slip the bonds of domestic motherhood and find both emotional and geographical space on the open road, tinkering is most commonly associated with masculinity and a male-presenting identity. Sociologist Frank Nutch's work identifies the role of "gadget-man" in the marine laboratory. He claims that the term has been carried over from a "less gender-sensitive time," when the biological sciences were male dominated. However, he never heard a female scientist referred to in this manner, and all the "gadget-men" he met were male and assumed that title without question. Bess Williamson and Franz both highlight the way that tinkering gives agency and the ability to form community in a society where consumer goods often take autonomy and knowledge from the consumer. Knowing how to fix your car carries a form of masculinity to the tinkerer.[16]

The aquarium is a tinkerer's technology. By this I mean that the aquarium cannot be standardized enough to run effectively without consistent observation and adjustment by someone who understands the system. The aquarist tinkerer, the person that builds and maintains the aquarium, must know how the system operates and be able to make changes quickly to maintain the function of that equipment. Expertise about the aquarium is developed through successful tinkering; even in the scientific laboratory, the "gadget-man" who tinkers with the system is one that works creatively using a wide range of knowledge brought to bear from a wide variety of issues. It is not uncommon to find each tinkerer to be self-contained in their work. While speaking with a tank designer in an academic laboratory, I asked how he had developed a tank design for a set of experiments. The kreisel tank design, built to keep jellyfish alive, is one that is well known throughout the aquarist community, and basic schematics could be found online or in books. But this tinkerer did not use any of those sources. Instead, he used knowledge from other building projects, suggestions from the researcher that would be using the tank, and materials he already had in the workshop. When I asked how he developed the plan for the tanks, he stated "I designed and built my own house."[17] The tank design worked in a similar way to all tanks used for those organisms, but it was not built using basic schematics but using a tinkerer's tool kit.

Because of the tinkering requirements of keeping tanks, the aquarium has not become standardized even as it has become integral to marine-biological knowledge production. Scientific technologies become standardized to facilitate "coordination among actors involved in a technical system." It is through standardization of technology that science research is validated and shared throughout a global network.[18] These technologies become widespread, commercially produced, and standardized between laboratories around the world. We see this standardization in experimental systems that arose in cancer and genetic research in the twentieth century. The development of genetically modified rats made it possible for researchers to be working on the same organism and compare that research between labs. The standardization of the rats and the technologies of study surrounding them made it possible for researchers to believe that everyone was working under the same conditions with the same tools.[19] But the aquarium has resisted standardization and remains a tinkerer's technology even as they are integral to much of the work of knowing the ocean.

In the consumer world, the ability to tinker is often a sign of a simple or relatively new technology. The trajectory of tinkering with vehicles shows a decline from the earliest days, when basic auto forms could be modified for specialty purposes, to today's highly specific and intricate specialty vehicles, which resist tinkering and require craftsmen and engineers for fixing and modification.[20] But the aquarium is not a new technology. Why have aquariums then remained firmly in the realm of tinkering? One answer is that the components for a successful tank are so specific to that location and those system variables that even the simplest tank requires extensive tinkering. In today's hypercompetitive, corporate scientific atmosphere, the patenting of scientific technology has become part of the structure of scientific work. Many scientific breakthroughs come with an additional promise of financial control over the tools and techniques that accompany that knowledge.[21] But an aquarium system (a complete system including food, water, and tank) was not patented until Alex Andon's Jellyart system in 2013. As we will see in chapter 3, Andon's system requires tinkering for maintenance, and whole communities have arisen to help, making it difficult to see how Andon will protect his patent. Like all tanks, any tinkering to help the system function changes the original design. The standardization required for a patent is not achievable for aquarium systems.

In addition, tinkering knowledge is difficult to convey and often requires personal communication or personnel movement for knowledge exchange. Franz calls tinkering "fingertip" thinking; the nature of tinkering is a personal "feel" for the problem. In the marine world, those that are good at keeping aquariums are labeled "blue thumbs." However, the knowledge contained within those hands is not easily conveyed. Much of the craft knowledge is tacit in nature, meaning that it is difficult or impossible to transmit in formal writing. Tacit knowledge is defined as "knowledge or abilities that can be passed between scientists by personal contact but cannot be, or have not been, set out or passed on in formulae, diagrams, or verbal descriptions or instructions for actions."[22] Often, it is difficult to describe exactly what parameters are present in the success of a tank design or how one went about performing the work. For instance, in chapter 5, we will see that learning the process of breeding ornamental finfish often requires that aquarists travel to the laboratory in which that work is being performed. Merely speaking or reading about that process does not adequately convey the "fingertip" work or "feel" for the problem needed to learn how to duplicate those methods.[23]

Using a tank to understand the ocean means that it must function to maintain the captive ecosystem. For the most intricate systems, knowledge about tanks must be gathered from and shared with a wide community of people. The reliance on tinkering to successfully operate tanks and the tacit form that knowledge about tank operation takes necessitates a very open form of communication between a wide variety of aquarists. Hobbyists, public aquarists, and academic researchers all develop craft and tacit knowledge using tinkering, and they must pass their knowledge through a variety of interactions, including writing, interpersonal communication, and the movement of people. This type of tacit knowledge is known as "embodied skill" and requires contact to be transferred.[24] Information about keeping tank systems can be found in hobbyist magazines, academic articles, full-length manuscripts, and online forums. But those writings are only the first step to linking this community and understanding how knowledge travels. The next section will highlight the different communities that I talk about in this book and show the intricate and often meandering ways that the embodied skill of tank craft makes its way throughout the network of aquarium users.

Defining Communities

A large and diverse community is dedicated to the development of tank craft. In its simplest form, the aquarium is a financially accessible technology that can be purchased or built to specifications. As a true tinkering technology, the changes made to the basic aquarium structure are usually done by individuals hoping to solve a given problem with the tools available to them. This means that while a fisheries biologist and public aquarist working to outfit a tank for breeding fish might start with similar tanks and knowledge about the fish, their eventual systems will look different depending on the materials available in each location. Both communities are capable of building tanks to perform the functions required, and there is no guarantee that any particular design will be more widely applicable. Aquariums built by hobbyists can function as well or better than those built in academic research laboratories. Hobbyists, academic researchers, and public aquarists all contribute to the development of specialized tanks, and it is through the sharing of knowledge that those interested in tank development refine their techniques and tools.

The members of the tank craft community are all contributing to knowledge about the marine world. A hobbyist and a professional researcher might simultaneously be working on fixing the same issues involving breeding a sea urchin. The purpose of their work, and the structure that their knowledge takes, is different, but the process of tinkering is similar. Because it is difficult to standardize the aquarium, oceans under glass are similar to the open ocean; there are many localized environments. A hobbyist rearing a fish in a basement represents a different marine environment than the same fish reared in a laboratory on a college campus or in a large tank at a public aquarium. All three tanks are technically doing the same thing—rearing a fish—but the act of doing so in these different microclimates creates knowledge about how these fish survive and thrive in varying environments. Each groups' findings result in increased knowledge about their tanks and the marine world more widely.

Historians and sociologists of science have long tried to understand the relationship between different "ways of knowing": how do varying groups of people understand the natural world with different tools and knowledge systems? The answer to this question has often resulted in a division between academic science and those who participate as "amateur scientists" or "citizen scientists."[25] These divisions are drawn through recognition of communities that form around educational or professional status. But looking at the aquarium community allows us to see that the use of the aquarium is a "way of knowing" about the ocean that is spread throughout a wide network of users with varying levels of educational and professional ties. This does not mean that there are not divisions but only that the knowledge created by tinkering allows the sharing of that craft that transcends those boundaries. Looking at the tank community shows that this network undergirds marine investigations; each member of the network contributes to marine knowledge. There is not a useful division of knowledge based on professional boundaries, and all the tank work discussed in this book—and the subsequent knowledge of the ocean derived from this work—is a combination of multiple communities' contributions.

In addition to participation in tank craft, there are other similarities between this wide community of aquarium users. The most notable, and possibly the most obvious, is a personal and physical connection to the marine world. Surfers, scuba divers, and angling communities develop group identity based on shared marine experiences instead of education or professional identities.[26] Aquarium users often belong to at least one of these other communities and consider their physical relationship with the ocean an integral factor in the knowledge they bring to their tank work. Akers's posts highlight the importance that she places on her physical links with the ocean over her life. Many aquarists hail from marine areas or spent considerable time near the ocean as children. In addition, many of the individuals I interview credit scuba diving as an important part of their relationship with the marine environment. I can attest to this emphasis on diving as an organizing point in the community: the most common question my interviewees ask me is about my diving experiences. The physical experience of being in and around the marine environment is important to aquarists, and they consider their movements between the open ocean and oceans under glass to be integral to their craft work. This holds true throughout the history of aquarists. Other identities that one might assume draw the community together, such as a commitment to a certain form of marine conservation or environmental ethic, are more divisive and shared primarily in the subgroups of the network. Instead, the aquarist community is linked both through aquarium craft and these perceptions of themselves as linked to the wider ocean through experience and emotion.

While this network shares knowledge, they do have separate histories and coherent communities. Public aquarists, hobbyists, and academic researchers' use of aquariums arose in different locations and for different purposes. Each group developed separate reasons for using tanks, and, importantly, they have different standards for creating and acknowledging knowledge using these systems. As sociologists of science have shown, understanding how scientific communities develop and communicate can shed light on how they produce and substantiate knowledge claims.[27] However, tank craft travels through these groups through a combination of publications, human movement, and sharing of spaces. It is important to understand the separate nature of each group to respect the way its members share information to place aquariums under glass.

PUBLIC AQUARISTS

Public aquariums are spaces that display aquatic organisms to educate and promote conservation to the public. The term *public aquarium* may call to mind a space financed or run by state or federal governments, but the term *public* in this instance refers to the idea that anyone can visit the space, not that it is run as a public good. SeaWorld, a for-profit entertainment park, is a public aquarium, as are all three aquariums run by the state of North Carolina. Public aquarists at these institutions are paid professionals working to breed and maintain captive organisms for exhibition and conservation purposes.[28]

Public aquariums were built throughout the world beginning in the mid-nineteenth century. These spaces were initially dedicated to presenting a wondrous image of the underwater environment previously inaccessible to the public. Early aquariums such as those in Berlin, London, and the United States initially focused on educating the public on the flora and fauna of their national waters and their colonies.[29] Most tanks contained native species and explained their habitat, habits, and the proper ways to catch and prepare those that were edible. Interest was spurred by the inclusion of large exhibits at International Fisheries Expositions. The history of these expositions has not been told as enthusiastically as that of World's Fairs, but they were very similar. Nations sent large displays of aquatic goods and equipment to show their wealth and power. The largest of these expositions took place in Berlin in 1880.[30] In addition to education about natural history and national resources, public aquariums were invested in early conservation initiatives. The earliest American aquariums grew out of Fisheries Expositions and took their mission directly from national fisheries policy. They stocked many of their tanks with local fish species that were then reared and released in stocking initiatives. Aquariums worked with fisheries organizations and conservation groups to try to breed rapidly diminishing fish, reptiles, and amphibians.[31]

Over time, exhibitions at public aquariums expanded to replicate the wider ocean under glass. Most major aquariums now exhibit not only local but exotic species sourced throughout the world. At the Aquarium of the Pacific in Long Beach, California, the coastal exhibits show local species, but there are also exhibits for open ocean, tropical seas (including Bali), and the Amazon. While there are still small aquariums dedicated entirely to local environments, it is common for public aquariums to show visitors multiple ecosystems and to educate the public on global understandings of conservation. This expansion of exhibits has also been met with an expansion of conservation initiatives. Today, aquariums breed endangered species, stock local waters, fund basic research, and are the most trusted place for the public to learn about ocean conservation and climate change.[32]

Aquarists perform many functions at public aquariums. The most basic functions of an aquarist are to monitor and maintain aquatic exhibits. This includes feeding, treating for basic illnesses (mites and bacterial infections), and monitoring the tank inhabitants for behavior shifts. While aquarists can get basic understandings of their charges from books and other resources, every tank requires continuous monitoring and tinkering because of local conditions. In addition to this basic care regimen, aquarists can also specialize. Husbandry experts not only care for animals but also work to breed those animals in captivity. Species Survival Plans (SSP)—a program between major aquariums to breed endangered species in captivity—require the development of husbandry techniques for endangered specimens, a process that can take years to perfect (if ever). In addition to endangered species, aquarists breed common species to decrease collection, and stress, on wild populations. Other aquarists specialize in enrichment—the development of techniques to maintain animal welfare in captivity. Many higher-order species, including octopuses and mammals, require mental and physical stimulation such as puzzles or toys.[33] In small aquariums, aquarists do all of these activities as part of their job description, but in larger institutions there are more specialized professional roles.

The aquarist community is bound together by organizations and publications. The American Association of Zoos and Aquariums (AZA), founded in 1924, links members into a network through publications, knowledge and resource exchange, and standards of practice in tank craft. To be a member of the AZA, an aquarium must meet high standards of animal welfare and dedicate the majority of their resources to education, research, and conservation. The World Association of Zoos and Aquariums (WAZA), founded in 1935, is the umbrella organization for the community with regional organizations, such as the AZA, on every continent. These groups share conservation goals and work to enact those plans. Species Survival Plans are written plans shared throughout the AZA network that detail the threat to endangered species, the goal of aquariums in the survival of that species, and the plan to meet that goal. For many species, the threat is human incursion on environment and the loss of species diversity. The goal is for aquariums to maintain genetically diverse captive populations of those endangered species and eventually release those individuals into the wild to strengthen existing populations. To enact an SSP, aquariums must work together to share information and techniques and trade individuals throughout the network to maintain genetic diversity.[34]

Aquarists communicate their craft primarily through interpersonal communications at workshops, conferences, and informal word of mouth. Aquarists do publish their work in academic journals, but these publications are not necessarily geared toward other aquarists but toward a wider specialized community of scholars, many of whom are not public aquarists but academic researchers. There are few journals specifically dedicated to the public aquarist craft. The *Drum and Croaker*, a journal dedicated to public aquarists, began publication in 1958. According to its history page, "Drum and Croaker is currently published annually, but prior to 1985 was published on an erratic schedule of zero to four issues per year, depending on the whims of the editor or apathy of the contributors."[35] The majority of their communication, especially surrounding tank craft, is done in person. Yearly workshops and conferences provide space for hands-on learning from those who have succeeded in their craft. For instance, the Monterey Bay Aquarium has been a leader in jellyfish husbandry since the 1980s (chap. 3). However, there are very few journal articles or formal writings on their techniques. Instead, aquarists with jellyfish skills spread their craft knowledge through a yearly "jelly school" that trains aquarists in those techniques.[36]

Recently, the Covid pandemic laid bare this reliance on face-to-face communication for craft exchange. In the spring of 2020, Piscine Aquatics, a commercial aquarium fish food producer, hosted a series of workshops titled Quarantining with Curators that invited aquarists to talk about tank craft with each other. These online meetings covered everything from coral husbandry to enrichment activities to combat the inactivity of animals in lockdown. The presentations were geared specifically to professional aquarists; participants were solicited through private communication and professional Facebook groups. Upon signing up for each session, participants were asked to identify their institution of origin. All discussions started with the understanding that the audience was familiar with professional aquarium functions and that they were there to exchange technical information about maintaining specific species in captivity. Each conversation was attended by a wide range of aquarists from throughout the world.[37]

ACADEMIC RESEARCHERS

The academic researchers I highlight in this book are experimentalists and observationists using tanks to supplement field research. The tradition of using tanks in the laboratory began in the mid-nineteenth century with the rise of German-based experimental biology. A rash of marine laboratories, built at the seashore, arose throughout the world between 1850 and 1930. These spaces offered researchers a chance to use plentiful organisms, from simple plankton to fish species available in many locations. During this period it was difficult to transport these delicate organisms inland, and tank craft was not evolved enough to maintain them inland for long periods. Biologists had to travel to marine laboratories to work with fresh organisms in controlled conditions.[38]

The most important tools in these spaces were movable aquariums with running water, which were able to be shifted from shoreline to laboratory to fulfill the requirements of a wide range of scientific inquiries. Marine laboratories attracted a diverse range of researchers using a broad range of organisms for their research. One researcher might be studying jellyfish reactions to nerve stimulation next to another studying how oysters filter feed. But everyone using marine organisms used early tanks to achieve conditions comfortable enough that their experimental subjects could be observed and experimented on. These laboratories became important crucibles for tank craft development in academic biology, especially surrounding model organisms such as jellyfish (chap. 3). And as that tank craft became more stable, it spread to inland laboratories and other scientific spaces.

Academic researchers demonstrate expertise by publication in peer-reviewed journals. Publications dedicated to marine research were limited to those published by marine stations until the middle of the twentieth century. If researchers worked at a particular marine station, they could publish their work in those proceedings. However, the readership of these publications was very limited. Most researchers working with tanks published in general science journals, but these publications do not contain as much information on tank systems and the tinkering involved. For instance, in chapter 3, we see the work that was done at the Plymouth Marine Station by Edward Browne to maintain jellyfish in a specialized tank. His initial tank design was published in the *Journal of the Marine Biological Association of the United Kingdom*, the journal of the marine laboratory. But subsequent work done with Browne's jar—but with limited discussion of the parameters of the tank—was published in physiology and biology journals without a specific marine focus. *Marine Biology*, the premier academic peer-reviewed journal for "life in ocean and coastal waters" began publishing in 1968. Journals focusing on specific marine topics, including mammalogy and coral reefs, were also founded during this era. The first volumes of *Aquatic Mammals* and *Coral Reefs* were published in 1972 and 1982, respectively. Today, there are a variety of specialty journals in which academic researchers and aquarists publish; aquarists often publish within the academic sphere, but academic biologists rarely publish in journals dedicated to public aquarists.

Tracing tank craft in academic research is difficult because of professional stratification in academic laboratories. Academics use tanks in their research, but they very often fail to fully report the variables used to maintain their specimens in captivity. For instance, it is rare to read a paper in an academic journal that gives full dimensions of a tank, logistics of the filtration system, water chemistry, food and temperature variables, and lighting parameters. Unless a publication is testing one of these specific variables, they are often only lightly touched on. One reason for these omissions might be that the process of keeping tanks in laboratories often falls to younger laboratory technicians, many of whom learn the craft from other technicians during their training and daily work. This training includes tacit knowledge that is rarely written down, and if problems arise, technicians commonly tinker with the system until they reach a successful outcome. This knowledge is then passed to the rest of the laboratory but rarely to the larger academic community via formal publications. While many academic biologists work with tanks intermittently, they leave much of the tank craft to lab techs and undergraduate workers.[39]

HOBBYISTS

Hobbyists are not a professional group. Instead, they are a community of individuals held together through a set of common practices and goals. A wide range of individuals operate home aquariums, but only a small subset of that community fall into the group of hobbyists that I speak about in this book. So what is the distinction?

The home aquarium emerged as a system with a variety of uses. The history of keeping aquatic organisms in captivity is ancient; throughout the Mediterranean and Asia, archaeologists have found evidence that a variety of fish and invertebrates were held and bred for food and aesthetics. The glass-walled home aquarium was introduced in Europe and the United States in the middle of the nineteenth century. The falling price of plate glass combined with new biological ideals about balanced ecological systems resulted in the development of the "balanced aquarium"—a self-perpetuating system of plants and animals that needed minimal care. Amateur naturalists used aquariums to collect and study marine specimens, and educators used the aquarium to teach both biological and religious beliefs. The "balanced" aquarium was used to teach students how species evolved to exist in a specific ecosystem. It was also used to demonstrate the nature of God's perfect creation, with everything in its place.[40]

The home aquarium craze also gave rise to a community of users interested in pushing the limits of captive breeding. The first fish fancying began in China around AD 960 with a focus on the carp and resulted in the development of the goldfish. Chinese goldfish fanciers exported them to Japan around AD 1500 and Europe two hundred years later. Fanciers bred goldfish with new coloration, longer fins, exaggerated eyes, and a variety of head and body modifications.[41] Goldfish fancying in the United States became popular in the late nineteenth century, and a firm in New Jersey supplied thousands of Chinese varieties to American hobbyists along the East Coast. By 1894, two firms, one in San Francisco and one in Seattle, imported goldfish from Japan for the American market.[42] But the process of fish fancying found fertile ground and a strong tradition of animal fancying in the middle classes—breeding cattle, dogs, and cats was a long-standing hobby in Europe and the United States, and fish breeding was not necessarily different. The process of fancying is a social affair—a successful breeding experiment is only completed when shared with a community that acknowledges that success. Fish fancying led to the building of hobbyist communities.[43]

As home aquariums became popular in the United States, aquarium societies formed in major cities. By the turn of the twentieth century, Philadelphia, New York, Boston, and San Francisco had aquarium societies with regular meetings, paid memberships, and newsletters. These groups met in public aquariums and meeting halls once a month to display their fish, award prizes for superlative specimens, exchange information about breeding techniques, and trade breeding stock in the hopes of copying or even inventing new varieties. Aquarium societies were initially male bastions, following a trend in which middle-class white males transitioned from belonging to men's clubs and groups into other forms of leisure.[44] Eventually these communities began accepting women, although the hobbyist community is still heavily skewed toward male participants.

The growth of the aquarium hobby resulted in national publications and conferences that turned a highly local hobby into an international community. William T. Innes published the first American magazine dedicated to aquarium hobbyists, *The Aquarium*, in 1932. Innes was a member of the Philadelphia hobbyist community and sought to link American hobbyists with a magazine dedicated to highlighting advancements in the field through essays accompanied by color photographs (chap. 2). Innes's magazine was for the generalist, but by the time it stopped publication in 1967, there were a variety of other publications that drew together highly specialized communities, including those interested in specific species or types of tank systems. Local communities eventually expanded into a national and international network of hobbyists. In 1973, the Federation of American Aquarium Societies (FAAS) was founded as a way to push back against proposed bans on foreign fish imports. The FAAS promotes hobbyist communities in North, Central, and South America. In addition to generalized aquarium societies, specialized groups such as the Marine Aquarium Society of North America (MASNA), founded in 1988, brought together national and international groups dedicated to specific aspects of tank craft.[45]

The development of national and international conferences further strengthened the links between local groups. Shortly after the founding of MASNA, the Marine Aquarium Conference of North America (MACNA) convened in Toronto and attracted 190 participants from Canada and the United States. The 2018 MACNA in Las Vegas, Nevada, attracted 6,277 aquarists, just seventy-three tickets shy of the largest MACNA attendance in Washington, DC, in 2015. These large conferences bring together far-flung members of the hobbyist community to listen to lectures, meet vendors, and exchange information with others. The explosion of conferences, magazines, and social media has created a tight-knit, continuously conversing community of hobbyists.

The Tanks That Bind

Each group of aquarium users is embedded in a network that creates and spreads knowledge. Within each group, advancement from novice to expert relies on participation in activities conferred with value based on community standards. In the academic research community, degree completion and publication in peer-reviewed journals are considered demonstrations of knowledge. Public aquarists advance in their profession through the demonstration of prowess via new exhibitions, publications, and lectures. And hobbyists publish in trade journals, display working systems, and communicate their work at club meetings and online. All of these activities that confer expertise have a history in the groups, and as methods of communication and research shift over time, new areas of expertise and avenues of acknowledgment are built into these groups. But creating marine knowledge with tanks relies on the combination of knowledge from these groups. So how does tank craft travel between them?

Publications are one way that information travels. As we have seen, each subgroup has agreed on publication standards that work to show expertise in that community. But it is difficult to trace the movement of information by publications alone. The most formal publications, hobbyist manuals and academic journals, have drastically different methods of citations and attribution of knowledge. This started particularly early. After the first issue of *The Aquarium* was released, George Sprague, an academic researcher and the secretary of the American Society of Ichthyologists and Herpetologists, wrote to William Innes to congratulate him on the magazine and to offer advice to improve the magazine. According the Sprague, one major problem was that Innes did not include bylines and more citation of materials for scientific readers. While hobbyists made up the majority of readers, Sprague stated that "scientific readers are going to have to refer to your magazine (as to other aquarium magazines) for accounts of the habits of fish, etc. to be found no-where else. Certain things *can* be done which will make it easier for these people to use the data you publish without inconveniencing your aquarist readers."[46] While Innes took many of Sprague's suggestions, it isn't clear that scientists who read these publications actually cited them consistently. Instead, they often used these sources as a starting point and either confirmed much of the data in their own experiments or took the information as general knowledge and therefore not required to be cited.

Today, citations in popular hobbyist magazines and books are mixed; hobbyists cite in their work, but this practice is not consistent and depends more on the subfield and community than on general patterns or rules. Many hobbyist manuals contain some citations to other manuals and academic journal articles, but they are in no way exhaustive in their citations. In Chad L. Widmer's *How to Keep Jellyfish in Aquariums* (see chap. 3), there are a handful of citations at the end of each chapter, but Freya Sommer, the former head of jellyfish husbandry at the Monterey Bay Aquarium and the person whose work jump-started the husbandry of most of the species in the book, is not cited even though she has several published papers. In chapter 4 we will see that reef tank hobbyists are especially good at citing their predecessors in both hobbyist and academic circles, but they are by far the most conscientious of the hobbyists for citations. In academic peer-reviewed journal articles, citations are often limited to studies that meet academic criteria of truth claims. This largely eliminates citations of hobbyists' or aquarists' work published in nonacademic publications. More commonly, academics place comments about nonacademic work in their acknowledgments. While they might learn from Mary Akers's blog, it would be rare for them to cite that information in their publications.

This inconsistency in citations does not mean that information is not traveling, only that it also has another mode of transmission. Tank craft travels primarily through human circulation. While groups that work with aquariums have varying ways of knowing, the aquarium itself provides the ability for individuals from one group to interact with and move into another group. In this way, the aquarium functions as what historians Susan Leigh Star and James Griesemer have termed a *boundary object*. According to Star and Griesemer, "boundary objects are objects which are both plastic enough to adapt to local needs and the constraints of the several parties employing them, yet robust enough to maintain a common identity across sites. . . . The creation and management of boundary objects is a key process in developing and maintaining coherence across intersecting social worlds."[47] In each group using aquariums, the tank is used for different purposes, and the knowledge created from them takes on different forms. However, someone who has perfected a tinkering process in a home tank can work with a laboratory tank and expect to achieve a similar outcome. The consistency of the aquarium's basic form means that the technology is legible to the aquarist regardless of the location and goal of its use. In the course of their lifetime, an aquarist might move between separate groups, with only the aquarium as a consistent tool in each space. It is not uncommon for an individual to be a professional aquarist or academic researcher and participate in the hobbyist community.

Martin Moe's career gives a good example of the movement of knowledge through human circulation. Moe began his career as a research scientist at the Florida State Marine Research Laboratory in 1962. During his tenure with Florida Fish and Wildlife, he began work on breeding marine species and eventually succeeded with clownfish. This work on clownfish was done as a hobbyist—it was not a part of his official duties for work. He then used his technique to begin an extensive career as a commercial fish breeder. Throughout the 1980s, Moe wrote some of the most influential hobbyist manuals and continued to push the limits of ornamental aquaculture (see chap. 5). Today, Moe is once again tinkering with breeding, this time the *diadema* urchin, to help revive Florida's imperiled reefs. Now retired, Moe is again a full-time hobbyist but one who has achieved fame in the community, having received the first lifetime achievement award from MACNA. He works closely with conservation groups, including academic fisheries groups at the University of Florida, and the hobbyist community. Throughout his career, Moe has been part of the academic, hobbyist, and research communities, often in overlapping ways.[48] Moe's career shows how information can travel through the movement of individuals, but it also highlights the difficulties inherent in trying to trace craft knowledge.

Tracing intragroup craft knowledge transfer, and tracing intergroup movement of that knowledge, requires a combination of historical and sociological methods. This book looks at the history of four tank designs that developed through collaboration between tank users from a wide range of professions and cultural groups. The earliest tanks began in the 1850s, and I have worked to trace the transfer of craft knowledge through a variety of historical methods. Each group of aquarists has publications that can be referenced. In addition, I used image analysis and archival research to piece together the working and social relationships of those sharing tank craft. Because much of the transfer of knowledge is person-to-person and informal, I have combined these traditional historical approaches with extensive interviews. I have sought to interview as many of the actors in this book as possible to understand how they learned their craft and to ask about their perceptions of the history of their work. One consistency in these interviews is that aquarists are often unable to answer how they know something and from whom they learned it. For instance, in one interview I asked how an aquarist learned a technique. The interviewee was unable to answer the question and seemed to point me toward someone who learned it from him. In this way, interviews have been useful for gathering historical data but also in understanding the nature of knowledge transfer as being nonlinear and difficult to trace.

Finally, to better understand the nature of the aquarium craft and those who work with tanks, I have adopted sociological methods of observation. In the last decade, I have taken tours and interviewed aquarists in research laboratories, public aquariums, and homes. In addition, I have observed daily work on tanks in several public aquariums and laboratories, including the St. Lucie County Aquarium in Fort Pierce, Florida, where I spent a week watching the operations of the aquarium and interviewing staff. These observations and interviews yielded historical information, and many were accompanied by archival research. However, they also allowed me to better understand how aquarists approach tank work and to observe the transfer of tacit knowledge, a type of knowledge that is, by definition, impossible for the user to acknowledge and discuss. Because I am not a member of the tank craft community, I do not take certain knowledge about tank keeping for granted; I lack any muscle memory of tank craft and therefore am able to pinpoint instances where the work witnessed cannot be explained by those I interview. This has made these observations especially fruitful, and I believe that the combination of historical and sociological methods has enhanced my understanding of the nature of knowledge creation and historical transfer in tank craft.

Learning to See: Oceans under Glass as Imagined Worlds

The history of tanks can tell us much about how we imagine marine spaces. For many people, their first and sometimes only interaction with the submarine environment is through aquarium glass. But the aquarium is not an exact replica of the marine world but instead contains the objects that we want to or can keep in captivity. This selection of variables creates an imaginative environment that can influence the way that people *see* the submarine world. In 1951, William Beebe, one of the most famous early twentieth-century ocean explorers, wrote about his first experience using a diving helmet to explore the underwater world. Beebe worked at the Bronx Zoo and spent a good amount of time at the New York Aquarium preparing for and imagining his explorations. About his first dive with the helmet on a coral reef, Beebe states that "I felt only as if I were in a very small, strange, but perfectly comfortable room, looking upon a wonderful tank of living fish with a most excellently painted background." To snap himself out of his simulation and return to reality, Beebe sits on a rock and thinks to himself,

> I am not at home, nor near any city or people; I am far out in the Pacific on a desert island, sitting on the bottom of the ocean; I am deep down under the water in a place where no human being has ever been before; it is one of the greatest moments of my whole life; thousands of people would pay large sums, would forego much for five minutes of this.[49]

And according to Beebe, this is all it took. When he opened his eyes, he noticed a "red bull of Kim" blenny staring at him from a nearby rock. He describes the coloring and behavior of "his blenny" in detail and continues with his narrative, having been shaken out of his simulation and into reality. Beebe's confusion between oceans in the wild and oceans under glass points to the power of the aquarium to affect human interactions with the sea.

Aquariums belong to a group of technologies that allow the user to replicate environments in miniature through environmental engineering.[50] The simplest tanks contain only a few "essential properties" of the marine realm—the coordination of water and food capable of maintaining an organism for a limited time.[51] These boxed environments, engineered as containers for a set number of controllable variables, make it possible to experiment to better understand both the way the system functions as a whole and the importance of that system to the organisms contained therein. This form of environmental replication has been popular from the nineteenth century onward. From early zoological and botanical gardens and test agricultural farms to large-scale techno-environmental builds such as phytotrons, the building of these replicate environments has served as a way for researchers to try to understand complex ecological systems by reducing them to their essential aspects and combining them in controlled spaces.[52] The simple tanks contain only a few variables from the environment to control, but the larger ecosystems, such as the reef tanks in chapter 4, are meant to replicate an entire coral reef ecosystem. But similar to the reef ecosystems, the larger and more intricate the replication, the less can be controlled in any simulation run in that space. Biosphere 2, a large complex built in Arizona containing a variety of environments meant to replicate Earth in miniature, is a classic case study of the difficulties involved in controlling variables in intricate engineered environments.[53]

The tank as a box can tell us much about how we imagine the marine environment and what we want to know about it. Scholarship on boxes suggests that understanding the container and what it contains helps us understand how humans have sought to order the world. The aquarium, at its core, is a box that holds imagined environments questioning and confirming ideas about animals and their relationships to both human and natural environments. At the turn of the twentieth century, natural history displays in museums shifted to the new "biological group." This type of diorama depicted a group of animals, as family units, in an environment. The grouping was meant to convey both animal behavior and the relationship of animals to the environment in which they were depicted. These biological groups were extremely popular with public audiences but were seen by some scientists as presenting tenuous understandings of behavior. Today, we know that many of the earliest dioramas and environmental representations were based more on human beliefs of how animals should behave and less on how they actually behave. In this way, boxes "shelter within themselves both order and materiality, thereby producing our knowledge about the world."[54] The box as built, as well as the builder, constrains the environment contained therein.

Marine dioramas, the preface to aquariums, can tell us much about the goals of the builders. One of the earliest displays of marine-biological groups occurred in the Museum of Marine Studies in Berlin in 1906. Alcoholariums, museum dioramas containing marine environments preserved in alcohol, presented audiences with marine worlds they had never encountered. Among many tanks were scenes depicting coral reefs and coastal spaces. These scenes served multiple purposes: to entertain a new scientifically interested public and to engage them in the new German goal of naval and oceanographic expansion. These tanks taught new ecological and evolutionary ideas but were also meant to teach the view that humans studying and eventually utilizing marine environments was both normal and beneficial for the ocean.[55] These marine dioramas were often the first glimpse of the ocean by the public, and their depictions shaped the way wide swaths of the public understood the submarine environment. Underwater dioramas often depicted dramatic "life and death" scenes that played on fears of ocean predators, including sharks and octopuses. Diorama builders for the Chicago Field Museum planned a scene involving a group of sharks attacking an injured fish. The fish's blood could be seen "diffusing thinly" up through the water. This description was given to the museum before the Field Museum–Williamson undersea expedition ever got close to the sea; it was a concept that preceded and shaped the expedition's research. The preconceptions that inform the creation of these early images of the sea come from a variety of locations, including literature and art.[56]

Dioramas and aquariums modeling the submarine world are often meant to portray more than a literal mirroring of that environment. The inclusion of sharks in the diorama and the violent nature of the scene was meant to teach museum goers the "natural order of life" in the ocean, depicting the predator-prey relationships on a reef and the harsh reality of the food chain and "survival of the fittest."[57] Other combinations of organisms were meant to show the perfection and balance in nature set in motion by a benevolent creator.[58] And the message of these tanks was of control and power over that realm as granted to wealthy, middle-class, male hobbyists struggling at the turn of the twentieth century to assert patriarchal claims in the changing home.[59] Aquariums became a craze that allowed the public to construct understandings of the subaqueous environment and to imagine the ocean as merely an enlarged parlor tank.[60] Public and home tanks still reflect lessons meant to train viewers to see the ocean. The current climate crisis and the impact on the ocean has been mirrored in public aquarium tanks—where there are exhibits on bleaching, plastic pollution, and biodiversity loss meant to spur viewers into action.[61]

Beebe's confusion during his dive is a demonstration of how environmental replications can simultaneously mirror and obscure the marine world. In his work *Simulacra and Simulation*, French postmodern theorist Jean Baudrillard theorized that models and simulations become realistic and overtake and eventually become indistinguishable from the natural world.[62] Scholars have debated the role of the zoo in simulating the natural world and the possible effect of that simulation on the way humans interact with it. Scholars point out that the aquarium is a space that requires the removal of animals from the natural environment to create a simulation that eventually overtakes and becomes the new "natural."[63] When Beebe first descends, he is surrounded not by a Pacific reef, but by his imagination of that reef through the constant witnessing of the New York Aquarium exhibits. He sees, immediately, the things that he recognizes from the tank: familiar fish, coral, and rocks. But it isn't until he shakes himself out of the simulation that he sees the things he doesn't know are supposed to be there. The blenny he sees, which is "three inches away from his face," was initially rendered invisible because it was not expected; that lack of expectation rendered the object invisible. Freeing himself from that simulation allows Beebe to see the sea instead of projecting the simple replication outward.

Beebe's encounter tells us why it is important to think about and try to understand the influence of aquariums on marine exploration and research more completely. Scholars have noted that aquariums and underwater visualization have changed the course of biological thinking. Research shows that the rise of the aquarium was integral in the development of ecological thinking in both scientific and public communities. Tanks are used to more fully understand the marine world, but they can also affect the perception of what belongs in wild spaces and what can and should be saved. For instance, placing used oil rig structures in public aquariums may show the viewer a reality—the rigs-to-reef program allows oil companies to keep rigs underwater as artificial reef structures—but it also works to "naturalize" those technologies and convinces the public aquarium visitors that these human-made objects belong in the ocean.[64] Currently, aquarium building is limited by financial, technological, aesthetic, scientific, and ethical concerns. The tanks that are constructed are not exact replicas of the submarine environment but are instead representations of an environment using available and possible means. They are also one of the most important avenues for visualizing the marine environment. The aquarium, through the choice and arrangement of organisms, is the creation of an idea of the underwater realm with real consequences. Whether you are a visitor to a public aquarium, a hobbyist, or an academic researcher staring into a laboratory tank, that aquarium is providing a visual experience that shapes your perceptions of what is and should be in the marine world.

The extension of the aquarium through other mediums extends the simulations developed through tank design. The earliest underwater photographs were taken not in the ocean but through aquarium glass. Carefully constructed images of marine organisms and tableaux, such as those I discuss in chapter 2, appeared in popular publications long before they were accepted in scientific journals. The earliest photographs were taken of organisms that often remain stationary, such as lobsters and octopuses. These creatures showed up nicely on film, but they also gave audiences a narrative about the underwater realm. The octopus appeared in literature as a villain, or an organism to be feared. In *Toilers of the Sea*, Victor Hugo refers to the octopus as a "devil fish" and includes a battle between an octopus and the narrator.[65] In popular images of the octopus, it is often pictured as a slightly imposing or scary presence. Pictures of coral reefs did not just teach people about the way that coral reefs looked but also carried with them ideas about colonial identity and modernity.[66] Photographs of aquariums carry the simulations of replicated environments into popular imagination.

Film has also given tank models more influence. Scholars have studied the way that film has shaped perceptions of the marine environment. The use of film to extend the simulation of the aquarium means that most people's understanding of the marine environment and its inhabitants has been "structured as a form of image."[67] The dolphin show was an important turn in public perceptions of dolphins, but the more important aspect of this dimension was the distribution of these shows via television and movies. Today, the images of playful dolphins performing tricks at Marineland have been overtaken by a more naturalistic type of tank film.

Public aquariums offer twenty-four-hour viewing of otters and coral reefs (Aquarium of the Pacific's Pacific Tank). These streaming images allow the viewer to imagine themselves as a submerged observer of the daily activities of a natural environment. Each camera is fixed, meaning that tank residents wander in and out of a shot instead of the seemingly more artificial concept of the camera following them. The viewer is sitting and watching the reef's life cycle. In addition to a lack of focus on a single organism, these cameras also crop out the tank structure and do not allow the internet viewer to see the aquarium visitors lined outside of the tank on the other side.[68] In this way, the viewer online is able to forget that this is a tank and instead imagine that they are in the ocean. The aquarium is a technology that mimics film and television, as these spaces make it possible for the viewer to see ever moving frames of an animal.[69] Gregg Mitman notes that the goal of Marineland was to

> reconstruct nature through science and entertainment. Indeed, Marine Studios is best read as a movie. The design and location of the portholes present un-paralleled photographic opportunities. If we stand back from the tank wall, each porthole represents a frame in the filmstrip, freezing the animal at a point in time. But as we put our faces to the glass, we become part of the undersea world. The task of Marine Studios, as of the natural history film, is to create the illusion of reality. By allowing visitors to meet the natural object in reality, rather than staging scenes in front of the camera, Burden hoped to unveil the story of animal life that is latent within the real, a story concealed from the scientist but that only science could eventually reveal. His goal in the construction of Marine Studios was to make the observer feel as though he or she was a witness to the activity of life off the coast of Florida, 75 feet below the surface.[70]

Aquariums are important technologies for understanding the marine environment. They also train the user and viewer to understand and see the ocean in a particular way.

We need to do the work of Beebe's observation: to make visible the frame of the aquarium so that we can see how it shapes our understandings and interactions with the marine world. The aquarium is a powerful tool for studying the ocean, and it is also a link between the terrestrial and aquatic worlds. Those that develop tank craft create knowledge that allows humans to "be the ocean" in whatever iteration they choose that day. But it is important to remember that these spaces are constructions that can affect the way we see and interact with the marine world. As climate change ravages marine environments, we must understand one of the most important but hidden tools in our ongoing efforts to know it before it is inextricably changed. Aquariums are integral to our knowledge of the marine realm and will take on even more significance in our bid to save those species and spaces we have come to value.

The next chapter explores the development of the photographic aquarium. Examining this tank—built to facilitate better photography of the aquarium interior and to simultaneously make it easier to erase the aquarium frame from viewers—shows the tensions of the aquarium as both a tool for research and a frame for our underwater imaginations. The aquarium as a visualizing technology, and the photographic tank specifically, introduces us to the ways that tank craft is built tank by tank but spreads quickly through the exchange of knowledge and photographs. It is a perfect place to start to understand the networks that build these tanks and the importance of aquariums in our perceptions of the marine world.

2

Photography Tanks

Viewing Oceans under Glass

In 2011 a group of researchers collected what they thought might be a new species of wrasse on an expedition to the Marquesas Islands in French Polynesia. The group was composed of academic biologists from the Université de Perpignan in France and the National Museum of Natural History in the United States. Based on previous descriptions of the only wrasse species in this area, the team had good reason to believe that they had found a new species. They collected eleven fish in total (four males, four females, and three juveniles) and documented the morphological and molecular characteristics of these fish. After catching the fish, the researchers noted the location of the catch, photographed the specimens, and then preserved them in 96 percent ethanol alcohol. After the expedition, a clipping of a fin was used to extract DNA material for a genetic analysis. The combination of these methods of description—including the geographical, physical, and genetic characteristics of the fish—revealed it to be a new species of wrasse, and the team announced the discovery of *Macropharyngodon pakoko* in a 2014 article.[1]

The classification of species is a scientific specialty known as *taxonomy*. Taxonomy is one of the oldest forms of scientific practice. The description of species and the ordering of them was carried out in the earliest civilizations. Early natural history encyclopedias, including Pliny's *Naturalis historia* written around AD 77, contained descriptions and images of the known animal and plant world. Many of these images and descriptions were based on word-of-mouth reports, and the artist drawing the object had never seen them. The images and descriptions contained in encyclopedias presented an abundance of species but were inconsistent about the important characteristics of each species that could be compared over time and space. Natural historians sought to impose structure on taxonomy, but it was not until Carl Linnaeus published *Systema naturae* in 1735 that a process appeared to unite those varying systems of natural history. Linnaeus divided the animal and plant kingdoms and used specific morphological characteristics to identify known or new species. For instance, plant identification became predicated on counting the sexual characteristics (stamen and pistil) of each plant. To help with the comparison of species, those adhering to the Linnaean system, known as binomial nomenclature, developed a variety of tools for taxonomy.[2]

Visualization tools are central to taxonomic research. Survey expeditions collect specimens and preserve them to compare later. Large animals are dressed and their skins kept for later examination; small species, such as insects, are dried and pinned on boards; plants are dried in herbariums; and fish and other aquatic specimens are preserved in alcohol. Natural history museums are repositories for these taxonomic collections.[3] But the physical specimen is not all that is required for taxonomic work. Since the spread of natural history collecting in the seventeenth century, when an organism is collected, a visual is also developed. Naturalists in the seventeenth and early eighteenth centuries included a wide variety of illustrations in their work. These images were not necessarily accurate morphological replications; they meant to emphasize and sometimes embellish the written descriptions, but they were not meant to help others compare species.[4] However, by the late eighteenth and early nineteenth centuries, new forms of species identification placed greater emphasis on the role of accompanying images. Linnaean taxonomy focused on specific morphological details to classify species.

As this taxonomic system spread, illustrations changed. Details such as the coiling direction of a snail's shell became important to portray accurately because the directionality of shell rotation helped identification. The spread of this system helped formalize images accompanying taxonomic descriptions. In the case of aquatic organisms, creating images of a specimen is an especially important part of taxonomic practice. When an aquatic organism is collected, it loses much of its coloration immediately upon death. When introduced to alcohol for preservation, some gelatinous species dissolve, and finfish usually turn an even shade of gray. To accompany the preserved specimen, the field images detailing the coloring provide important morphological information that can help identify the species. In addition, these images are later circulated to the taxonomic community to provide descriptions of the species because the preserved physical specimens cannot travel so easily.[5]

Early marine taxonomy relied on artist-produced illustrations to accompany species description. As soon as a fish was taken out of its native environment, a natural historian took notes on its coloring and other physical features. Then it would be placed in alcohol, and when the expedition docked, the notes and specimen would be provided to an artist or draftsman to be drawn for publication. Sometimes, expeditions hired artists to accompany them to draw the specimens while alive. Isabel Cooper spent seven years in the field working as an artist for the New York Zoological Society. She describes the moment that a fish is dropped into a tank for her to draw it.

> But it was a great mistake to spend much time upon reflection, because rage and discomfort had a strange effect on their color schemes. Right before my eyes the gleaming steel and gunmetal of their visors and armored plates would dim and darken and film over with streamers of purple mist, or jagged patterns of ultramarine, or shadows of leaden grayness. And I would be left guessing, somewhere between the myth of what they had been and the myth of what they were rapidly becoming, with nothing remaining of the truth which the scientists most earnestly desire.[6]

These initial drawings would eventually be refined and published to accompany descriptions of new species. The tradition of taxonomic drawings dates back hundreds of years. However, in the middle of the nineteenth century, a new form of documentation, the photograph, emerged.

New glass-making technologies made marine photography possible. In the middle of the nineteenth century, new methods of working, increased production, and falling prices pushed plate glass from a luxury to a commonplace object. One result was the development of photography. Cameras relied on glass plates to capture exposures, and the drop in prices made it possible to extend the use of cameras into new areas. But even cheaper plates could not make cameras waterproof—exposures required large flashes to capture images—and so marine photography also required the development of another glass structure: the aquarium. Early aquariums were expensive, and the glass was often thick or curved. The development of cheap methods for producing high-quality plate glass resulted in thinner, flat-sided aquariums. Cheaper photography technology and mass-produced aquariums eventually met when photographic hobbyists began trying to capture marine images in tanks. But these early photographic experiments exposed the difficulties of shooting images through glass and water.[7]

Producing usable photographs for marine taxonomy required tinkering with both tank design and photographic techniques. Early marine photographs did not meet the exacting standards required for taxonomic illustration. The process of shooting moving organisms through both glass and water proved to be difficult. The earliest aquarium photographers were photographic hobbyists working to push their hobby further and to show peers that they could do something deemed nearly impossible. They refined tanks and photographic techniques, and over time they learned to pair them for clearer images of marine subjects. In the late 1800s the earliest tank photographers could produce blurry images of largely stationary creatures (such as lobsters or octopuses). By the 1940s tinkerers were producing clear images of living fishes. It was in this period that the academic science community began to use established photographic methods and build on them to incorporate photography into formal aspects of scientific illustration.

In 2011 the team that discovered the new species of wrasse used photographic tanks and techniques to produce images of the fish to be used both in taxonomic identification and in their publications. Jeff Williams, an ichthyologist at the National Museum of Natural History, is a fish photographer. Williams brings glass plates with him on expeditions and uses marine cement to build a quick photographic aquarium on board. He props the tank on top of a sheet-covered box and hangs another sheet behind the tank to produce a white background. The fish are laid out gently after they are caught and lacquered to hold their fins in place. Then, Williams places them in the water-filled tank and pushes them flush on the front of the glass using a final piece of plate glass held in place with clothes pins. His method, a result of tinkering by photographic hobbyists and field researchers over the course of almost 150 years, results in images that show the morphological definitions of the fish important for identification.[8]

While the final images of the fish show the pertinent taxonomic details, they also hide the tank craft Williams uses to capture those images. If you did not know that these images were taken with a tank, there are no visual cues to explain the process used. Williams's tank design and his use of background color are his own. He has never published on his craft, and the process is not explained in any of the methods sections of taxonomic papers. Williams also struggles to pinpoint how he himself learned these methods. Photographic tank craft is an ephemeral history, difficult to trace and hard to spot in the final product. But in its near invisibility, it is also effective because it is an aquarium craft that reaches wide audiences: it shapes the way that viewers imagine the sea.[9]

Photographic tanks combine two different visualization technologies to produce a long-lasting and far-reaching oceanic vision. The aquarium is, at its core, a tool for realizing an imagined submarine world. The introduction of plate glass in the Victorian era allowed the viewer to take in an entire idea by gazing into a contained glass box. The inclusion of a variety of elements in a tank presents to the viewer a seemingly complete image of the underwater environment. Those studying the history of museum

FIGURE 1. Male (top) and female (bottom) color patterns of *Macropharyngodon pakoko*: **top** USNM 409153, 72 mm SL, male, holotype; **bottom** USNM 409154, 60.0 mm SL, female, paratype. Photographs by Jeffrey T. Williams.

FIGURE 2.1 Image of a new species of wrasse showing male and female specimens taken in a photographic tank onboard ship during an expedition. The tank frame disappears, emphasizing the color and morphological features of the fish. Used with permission of Jeffrey Williams.

dioramas highlight the way that the inclusion or exclusion of certain elements tells the viewer a story about natural history.[10] For instance, dioramas of human evolution often present images of heterosexual couples with the male leading a family through the landscape. This diorama does not just present early *Homo sapiens* in their imagined environment but also subliminally impresses on the viewer that the nuclear family is a recognizable evolutionary adaptation and therefore both historical and natural.[11] Early marine dioramas were meant to invest the audience in the new science of marine exploration and to depict the ocean as a scary and exotic place full of large predators, such as sharks.[12]

Others have compared the aquarium to a movie. In large tanks at public aquariums, the viewer's eye is guided to the center of the tank by the construction of frames around it. Viewers peer into the glass and see a moving image presented to them. Similar to dioramas, this moving image is curated to appeal both to the viewer and to introduce a narrative about the marine world. Large tanks usually contain a balance of predator and prey fish, but the viewer never sees the relationship between these animals acted out. Predators, such as small sharks or large fish, are fed regularly to reduce predation on other tank inhabitants. Instead, the viewer is offered a utopian view of organisms endlessly circling a coral reef, safe in an environment that contains both predator and prey but does not ever require them to fulfill their roles. Wildlife movies are characterized by "the visual splendor of pristine nature, charismatic mega-fauna, an avoidance of the topics of science, politics, or conservation, no traces of human civilization, a sense of timelessness, and dramatic or suspenseful storylines," and aquariums seem to fit into this definition as well.[13]

Photographic tanks offer the viewer a simple vision of oceanic imagination. Early photographic tanks contained several elements of an imagined ecosystem, including multiple organisms and plants. As hobbyists refined their technology and techniques, they zeroed in on certain plants and animals that photographed well, including octopuses and other slow-moving creatures. Academic researchers who took up the use of the photographic tank for their taxonomic work developed techniques that shaped the images produced. Taxonomic photographs stripped out the plants and multiple organisms to focus on the specimen being presented. But those photographers still made technical choices that changed the photograph. Of particular importance is background color. Some taxonomic photographers use black backgrounds and others use light. While they explain their choice based on the way the color shows taxonomic detail, their decisions can also affect the way the viewer sees the marine realm. Black backgrounds give the perception of depth and the deep ocean, and white or light background coloring shows the fish as floating in empty space. Even minor changes in technique affect marine imagination for those who see the images.

Historians and sociologists of science have highlighted the tensions present in the use of photography in academic knowledge production and dissemination. Early tank photography was often blurry, and few would suggest that it could take the place of illustration in academic work. In addition, many of the images were of posed organisms in environments from which they did not originate, marking them as something between art and science. These photographs were similar to others during the Victorian era that were seen as entertaining but not trusted in science. But as photographic techniques advanced, so did photography's place in academic research.[14] Faster lenses and shutter speeds and clearer photographs captured images that many felt could not possibly be seen by the human eye. These advancements led viewers to believe in the mechanical objectivity of the camera: photographs could unveil truths hidden to researchers.[15] According to Phillip Prodger, "Photographs assumed a dual role. They illustrated something, but they were also experiments in their own right. They became more than mere pictures—they became data."[16] But many still feared the artistic nature of photography. The importance of photographs as both illustrations and data forced researchers to become more thoughtful about where and how their photographs were created. Academic fields developed journals and technologies to standardize image production in the hopes of assuring mechanical objectivity.[17]

Marine taxonomic photography maintains this tension. Aquarium tanks require tinkering to develop localized techniques, and photographic tanks are no exception. Taxonomic photographers craft their tanks based on their locations, available equipment, and the organisms being photographed. A photographer working with an endangered species on a research vessel might build a tank meant to withstand rocking and that is conducive for live photography so as not to kill what could be one of the last of the species. A laboratory-based photographer trying to capture morphological features of a fairly common fish would build a very different tank. And both of these photographers would make different choices regarding background and lighting based on their research goals and their own aesthetic judgment. The loose standardization in tank photography is emphasized by the lack of publications dedicated to the craft. There are no journals for marine taxonomic photography and very few publications explaining the methods developed to produce these images.

This chapter will examine the development of the photographic tank craft. The history of this craft can tell us much about the diverse community of tank craft developers and how varying motivations for tank use result in shared knowledge. As might be expected, the first developers of photographic tanks were photographic hobbyists who also had an interest in the marine world. They began work with photographic tanks not only to create knowledge about the ocean but to raise their stature in the photography community by doing something no one had done before. Those individuals at the intersection of marine research and the photography community became knowledge couriers who circulated techniques throughout the marine network. Eventually more members of the network took up the camera and photographic tank in their work, but the impact and importance of the photography community continued, and the dual nature of marine photographs, as both artistic and scientific documents, persists today.

Exploring the history of photographic tanks can also tell us about the development and dissemination of marine simulations. Unlike tanks, which are relatively stable objects that do not circulate widely, images of tanks travel great distances and can have a greater influence on the way people imagine the underwater realm. In the simplest sense, photographs freeze a moment in a tank and make it mobile. For taxonomic purposes, the photograph that Williams produced of the new wrasse functions as a proxy specimen. If another researcher wishes to know whether they have found a new species of wrasse, they can compare their specimen to the photograph in the paper. Photographs allow researchers to access collections wherever they are and to compare and contrast their own specimens to those presented in those images. But photographs can spread not just information but ideas. Anne Elias has traced the life of movies and photographs made of Indo-Pacific coral reefs at the turn of the twentieth century. According to Elias, Frank Hurley shot many of his scenes for the movie *Pearls and Savages* using tanks. Her work shows how these carefully curated images of coral reefs circulated throughout the world, awakening a sense of modernity tied to visions of exotic locations full of Edenic coral reefs and primitive societies.[18] Photographic tanks train the viewer to see the submarine world as a simulation. To get clear photographs and images in tanks, photographers reduce the diversity of the ocean to a simplified scene containing only the focus of the piece. For instance, the BBC's recent success with documentaries such as *Blue Planet* and *Frozen Planet* have relied heavily on the perception of filming wildlife in its natural habitat. But many of the most fantastic images taken, especially in *Frozen Planet*, have been filmed in aquariums. For one shot, meant to depict the small life forms existing underneath melting ice, the film crew hoped to film in the field but encountered trouble.

> Filming small creatures is a huge challenge in any environment, requiring tiny specialist macro lenses focusing on minute subjects with complex lighting requirements. In this instance, the creatures were so small that the swell and current of the open ocean made them almost impossible for a camera team to follow and focus on. So, with guidance from scientists, the team recreated a little bit of melting ocean in a laboratory in Svalbard. Their specially built sea water aquarium featured a system to circulate the water keeping the delicate little sea creatures buoyant.[19]

Viewers of this tank craft would be unaware that the image they associated with the wild environment were merely the essential elements chosen to represent that environment in the laboratory. This simplified simulation affects the way the viewer imagines the underwater environment even as they fail to recognize it as a simulation.

The erasure of the aquarium frame in marine photography effectively interchanges oceans under glass with oceans outside glass and extends the simulated ocean. To capture the images of the Arctic plankton, the plankton had first to be transferred to a laboratory and settled into the tank. The decisions about how to build the tank and how much plankton to include were made by videographers compromising to get the best final image. The native environment of the plankton was impossible to capture in an aesthetically pleasing way for the audience; therefore, the image eventually captured was one that did not truly represent its natural counterpart. This process of building a tank for photography is one of compromise and aesthetics to find the best image for the outcome desired, whether it be entertainment or scientific claims. The decision to exclude or include certain species or objects to create a visually acceptable image is as much an aesthetic as a scientific decision. However, it is not clear that the audience for these images is aware of this process of omission and inclusion. It is important to reveal the layers of subjectivity in these supposedly objective images, especially those photographs purporting to capture a single specimen. Taxonomical photographs seem relatively straightforward upon first observation, but choices of backdrop color, fixatives, and tank and lighting design change the resulting image and can tell us much about the construction of subjective nature in these seemingly objective images. By understanding the development of photographic tools and methods, we can better understand the way the resulting images represent creative construction of the marine environmental imagination.

This chapter will examine the development of the photographic tank, including its technological and technical elements. This history sheds light on several major themes of this book. First, photographers from various communities developed the tools and techniques currently used throughout the marine network. The photographic tank was not initially developed by aquarium hobbyists but instead by photographic hobbyists to show their mettle to the larger hobbyist community. The intersection of the photographic hobbyist community with aquarium hobbyists brought the practice into the marine knowledge network, but photographic tanks and the continued development of new techniques to use them remains of interest primarily outside of the scientific community. The simplicity of the tank designs and the general availability of photographic equipment mean that the development and use of the photographic tank is one that encompasses a wide variety of users and producers of knowledge. Second, examining photography shows how wide-reaching aquarium visualizations can be. Both personal and public aquariums are simulated worlds shaped by their users. Examining these spaces can tell us much about how the aquatic environment is perceived. Combining our extended analysis of aquariums with images of those spaces adds another layer to our understanding of the construction of the marine environment in the mind's eye. Photographs of aquariums increase the exchange of a shared imagination of the submarine world.

Photographic aquariums require intricate and sometimes obfuscated forms of craft knowledge and produce lasting impressions of oceans under glass that can have very real impacts on human interactions with the submarine world. Tracing the history of the craft can reveal the architecture of the marine simulation created by oceans under glass.

Early Tank Photography

The development of the photographic tank demonstrates how important knowledge couriers are to spreading tank craft in the marine research network. Marine photography straddles two communities—photography hobbyists and marine researchers and hobbyists, and the photographic tank was initially developed by these individuals. Photographic hobbyists with interests in marine research brought expertise from the photographic community into the marine community. From there, the development of the photographic tank spread and entered academic marine science. But unlike many scientific tools used in academic settings, the photographic tank has not been standardized, nor is there a clear academic narrative regarding the development and dissemination of techniques and technologies. As we will see in the second half of the chapter, operating a photographic tank today still requires craft knowledge about technologies and techniques to be passed verbally through the tank network. The continued craft nature of one of the oldest specialized aquarium tanks shows how difficult it is to standardize oceans under glass and how important couriers are to the circulation of tank craft.

The earliest attempts at marine photography came from photography hobbyists with an interest, either professionally or personally, in marine science. Possibly the best known of the earliest marine photographers is Louis Boutan (1859–1934). Boutan was a marine researcher at Arago Laboratory in Banyuls-sur-Mer, France, who was interested in mariculture. He began wearing a diving suit to study in the wild mollusks that could not be maintained in the laboratory aquariums for the long term. After his initial dives, Boutan began to think about recording these dives for teaching and research purposes. Early photographers worried that water was not clear and would appear opaque in photographs. In addition, early photographs required long exposure times and continuous lighting. Working with his brother Auguste, Boutan eventually succeeded in taking some of the most iconic early underwater photographs.[20] His work with underwater photography settled questions of whether the technology could be used in that medium, but it did not produce scientific artifacts that could function as a form of field notation. His underwater portraits are iconic as early photography, but they did not become the pieces of scientific data that he wanted. During the same period, physiologist Étienne-Jules Marey (1830–1904) and early filmmakers and brothers Louis (1864–1948) and August (1862–1954) Lumière used aquariums to film the movements of marine organisms. These earliest experiments with underwater moving images had great influence on the cinematography community.[21] But it was another, lesser-known French photographer who developed the prototypical photographic tank and the earliest techniques for using still photography for marine study.

Paul Louis Fabre-Domergue (1861–1940) was from an old Creole family from the French island nation of Martinique. He studied medicine and physiology in Toulouse and did research on microscopy at the French marine station at Concarneau. After moving around for his studies and early medical career, Fabre-Domergue eventually returned to Concarneau as the director of the station. The station was dedicated to fisheries research and furthering understanding of French marine resources. Fabre-Domergue worked on sardine stocks and other fisheries concerns. However, he was also an avid photography enthusiast.[22] During his time as director, he sought to combine his day job of experimenting and exploring France's diminishing fisheries with his personal hobbies of art and photography. Fabre-Domergue was particularly interested in using laboratory technology to unveil the seemingly invisible natural world. In 1897 he published *Les invisibles*, detailing his observations of microbes using a microscope.[23]

Fabre-Domergue was one of the first photographers to successfully use the aquarium as a photographic frame. In his published portfolio *La photographie des animaux aquatiques* (1899), Fabre-Domergue hails the early aquatic photography done by both Boutan and Marey and seeks to extend those experiments. While Boutan produced underwater images, they had little clarity; Marey's work was impressive, but film was easier to take with moving marine organisms than still images. Fabre-Domergue sought to develop tools and methods for consistently clear, still underwater images. In the portfolio, he outlined his methods.[24]

To accomplish this feat, Fabre-Domergue built a specialized tank system. The tank itself was 70 × 50 centimeters (about 2.3 × 1.6 feet), with glass made by Saint-Gobain. The glass was specified as being of the highest quality available. Similar to the other aquariums in the laboratory, the tank had an overflow system and constant circulation to maintain water temperature and clarity. In the tank he placed rocks and plants in a small landscape from the original habitat of each species. To provide additional background and perspective, a second tank, containing more algae and organisms that were not meant to be the subject of the photograph, could be placed behind the first.[25] After the tank and landscape were built, a subject could be introduced well in advance of photographing. Fabre-Domergue suggested at least two hours so that the subject could get comfortable in the new environment.

In addition to the basic elements of the tank, Fabre-Domergue asserted that the most important aspect of aquatic photography is lighting. While one might be tempted to shoot in low light or direct sunlight, Fabre-Domergue strongly recommended shooting indoors with lighting provided by burning magnesium in the presence of potassium chlorate. This created a very bright light similar to a flare. While it provided extremely bright light, it was short lived, extremely hard on the eyes, and equally hard on the lungs. To use magnesium lighting, the photographer had to wait until the subjects settled into position. Then they quickly lit the magnesium in hopes that the tableaux would remain constant for the short period of extreme illumination. Fabre-Domergue built a hood over the aquarium and painted the inside white to reflect light back into the tank. The hood was open at the top, back, and into the aquarium and worked as a light tunnel, directing the harsh rays of the magnesium flares into the interior of the aquarium. During the photo shoot, this addition both funneled light into the interior of the tank and blocked the harshest lighting from the lens of the camera and the delicate plate being exposed.[26]

Fabre-Domergue also gave suggestions about technique: choose your subjects wisely. According to Fabre-Domergue, not every species was created equal when it came to sitting for portraits. Very quick fish and small species did not work well for aquatic photography; no amount of artificial light was strong enough to bring these subjects into focus. In addition, the sudden lighting required for exposure easily startled timid species. It was best to choose species that were comfortable in light and tranquil at the moment of shooting (his most famous portraits were of lobsters and octopuses). Finally, Fabre-Domergue's advice to photographers was to have patience; according to him, aquatic photographs required hundreds of exposures to get a single usable image.

Fabre-Domergue's biggest struggles, to find adequate lighting and to have patience, were mirrored by other early marine photographers. Aquariums, and especially marine aquariums, require a delicate balance of light to maintain organisms. Too much light and you get algae blooms and deoxygenated water; too little and you get fish with vitamin deficiencies. Each organism has a different requirement (some deepwater species don't need any lighting, whereas certain fishes need abundant illumination). Adding to the difficulty of finding lighting that worked for your organisms was the difficulty of making those subjects visible to an onlooker. For simplicity, the earliest large, long-term aquarium setups like those at public aquariums relied on natural overhead lighting and cave-like darkness for viewing. Most aquariums achieved this view by building grottos. The visitor walked through a dim hallway with aquariums set into the wall along the corridor. The aquariums appeared as paintings on the wall, and lighting was achieved by skylights above the tanks (but out of view of the visitor). This lighting and architectural scheme showed the aquarium interior to its best advantage to the viewer, and photographers wanted to replicate this in their images.

While Fabre-Domergue was working out his technique in France, an American serviceman, amateur naturalist, and photography enthusiast was doing the same in the United States. Captain R. W. Shufeldt (1850–1934) was trained as a medical officer in the US Army. During his active service, he worked as a surgeon during the Sioux campaigns in the western United States and developed an interest in natural history by collecting Native skulls to be sent back to DC for anthropological study. After his time in the Army, he became a curator at the Army Medical Museum (the institution holding most of his collections). His interest in natural history was largely shaped by his white nationalist and misogynist views, but he also showed a marked interest in collecting birds and in wildlife photography.[27]

In 1897 Shufeldt sought to demonstrate his prowess as a photographer by producing clear images of living fishes. It does not seem as if he had any special interest in fish or the marine environment but instead recognized the technical difficulty in capturing their image on film. In an article published in the *Bulletin of the United States Fish Commission*, Shufeldt points to Boutan's previous work (although, in true American fashion, butchers his name, calling him M. Bonton) and spends a significant portion of the article strengthening his claims that aquarium photography is extremely difficult and therefore his methods and successes should be taken as impressive. According to Shufeldt, the earliest images of fish were interesting but largely useless for scientific study. His goal was to produce photographs with enough clarity to serve as evidence in scientific research.

Shufeldt developed his technique at the US Fish Commission's aquarium in Washington, DC, at that time located in a building near what is now the Air and Space Museum. Between 1897 and 1899, he took thousands of photographs using a variety of tank and lighting setups. The first tanks he used were those in the aquarium grotto. In this space, there was natural light from above the tank, and Shufeldt set up his camera about two inches from the glass front. In July 1897 he produced several images of fish in these tanks, although he found that the light from above often washed out the fish in the final image. He blocked some of the light with umbrellas and suggested that photographers focus on fish that do not reflect light (catfish proved to be problematic because of their skin reflections). In addition to lighting issues, another problem turned out to be getting the camera prepped for the perfect moment. Instead of following fishes around the aquarium, Shufeldt focused on a space in the aquarium and set the shutter "snap": when a fish entered that space he quickly, but carefully, released the shutter to expose the plate.

One image produced during this period was that of a pike (*Lucius lucius*). Shufeldt describes his process like this:

> It had a duration of about 2 seconds, at which time the plane of the left side of the fish's body was nearly parallel to the plane of the glass, and about 3 inches from its inner surface. A quarter of an inch diaphragm was used, and the subject remained practically motionless during the time of exposure. Overhead the light was somewhat diffused, and an additional disadvantage presented itself in the fact that the color of the pike closely simulated the shade of the metal back of the aquarium, thus rendering strong outlines of the resulting negative matter in doubt. However, the picture (plate 7, lower figure) was fairly good, and on comparing it with the figure of this species in "The Fisheries and Fishery Industries of the United States" (plate 183, upper figure) it is to be observed that in the living fish, the pectoral fins are extended almost directly downward; and further, that the extremities of the forks of the tail are distinctly rounded and not acute, as in the aforesaid drawing. In fact, the caudal fin, or tail, in the latter is incorrect in outline and there are still other clear differences to be observed upon comparing the figure of the present paper with the figure given us by Goode, pointing to inaccuracies in the latter. Here is where the great value of the camera comes in. In time, with suitable subjects taken under the most favorable conditions, pictures of fish (as in the case of other animal forms), produced by half-toning processes from faultless lithographs, will surely supersede in biological literature the often inaccurate figures that now illustrate it. This is what we strive to accomplish in our efforts to obtain the best possible photographic negatives of live fish in their natural element, with normal surroundings.[28]

It is unclear in this passage whether Shufeldt considers his aquarium photographs as an example of capturing "the best possible photographic negatives of live fish in their natural element, with normal surroundings"

FIGURE 2.2 Photograph of a black bass by R. W. Shufeldt. Shufeldt sought to use photography to create scientific images that would provide more detail than those that were hand-drawn. However, many of the images included in his article, such as this one, were somewhat unfocused and showed the constraints of photography at this time. Bulletin of the U.S.F.C, 1899. Photo by R. W. Shufeldt.

or whether he considers the aquarium the first step in the evolution of fish photography with the eventual goal of tankless photography. Both Shufeldt and Fabre-Domergue call attention to the background of their photographs being natural, with plants and rock outcroppings that mimic the underwater environment. But in both cases, the photographers seem largely interested in placing their subjects at ease and suggest that providing a natural environment will help the subjects exhibit normal behavior (as opposed to stress responses). Both Fabre-Domergue and Shufeldt stated that calmer fish were easier to photograph because they were less likely to behave erratically. So the background, for the photographer, was used to calm the subject, not to portray scientific information about the environment from which the specimen came. However, the casual reference to "normal surroundings" suggests that, for Shufeldt at least, the photographers might have assumed that their backgrounds were obviously natural or normal. Fabre-Domergue was a marine researcher and had field experience with the marine environment, so he would have been aware of what plants and rocks existed near his subjects. But

Shufeldt, who had no marine experience, might have assumed that the background was natural because it matched his underwater imagination.

Shufeldt's vision of photographic images affecting scientific discourse did not come to pass in his lifetime. In fact, scientists proved reluctant to incorporate photographs into their academic publications. Shufeldt's images appear in popular publications, such as *National Geographic*. Arthur Radclyffe Dugmore, another pioneer in natural history photography, provided images using photographic tanks for David Starr Jordan and Barton Evermann's popular *American Food and Game Fishes* (1920). However, these photographs rarely appear without an accompanying line drawing of the same species, suggesting that even those researchers interested in the medium did not completely trust photographs to convey pertinent information for species identification. Aquarium photography and the development of photography tanks and techniques remained largely the purview of hobbyists until the mid-twentieth century.

Developing the Photographic Tank

The intersection of fish fancying and photography enthusiasts produced the next major developments in photographic tanks. The interest of hobbyists in photography stemmed largely from a community of fish fanciers working to develop new breeds of aquarium fishes (see also chap. 5 on breeding aquariums). Those breeders working to develop new physiological traits in fishes traditionally displayed them at regional hobbyist shows. For instance, a telescoping eye goldfish bred in Long Island would be shown at the New York City hobbyist show at the New York Aquarium. While this showing introduced the fish to the local community, it might not have an influence on the breeding community in San Francisco. Line drawings cemented species and variety standards, and many fanciers excelled at natural history illustration, but these images were distributed long after the fish's original showing and failed to advertise the breeding breakthroughs of particularly spectacular specimens. The development of photography in the aquarium hobby community facilitated the showing of these specimens nationally and internationally. The photograph became an eternal reminder of the perfection of an ephemeral offspring.

Imaging has always been an important aspect of breeding livestock and domestic animal fancying. Historians have shown that breeders commissioned images of prized livestock for themselves and to be circulated throughout the breeding community. These mementos served two purposes: to demonstrate and celebrate this one successful breeding venture and to fix that success as the new standard in the breeding community. Individual bulls had names and both physical and mental characteristics remembered by their breeders. But they also represented a yardstick by which other breeders measured their stocks and an idealization of the perfection that breeders constantly sought to reproduce.[29] These images were at once idealistic and scientific. Aquarium hobbyists used photography in a similar manner. They used photography to create records of individual breeding successes and to produce documents that would serve as models for other breeders. Here again, photographs served two purposes: display and standardization.

William T. Innes

At the center of the introduction of photography to the aquarium hobbyist community sits William T. Innes. Innes was born in Philadelphia, Pennsylvania, in 1874 and was a member of the Innes publishing family. His introduction to the aquarium hobbyist community came through his interest in photography. When asked how he came to the hobby, he wrote that "A friend at the Photographic society (Joe Powell) induced me to accompany him to the Aquarium Society in a 4th floor room at 1301 Arch St. (now long [torn] down), where I became an instant convert to the fancy goldfish and its enthusiastic breeders."[30] Although Innes does not give an exact date, this probably occurred somewhere around 1906 or 1907. Innes became an enthusiastic hobbyist and began breeding and showing his own fish. In 1908 he won awards from the Philadelphia Aquarium Society for his scaleless fringetail goldfish.[31] He continued to show goldfish into the 1920s. Eventually, Innes would become known as the "Dean of American Aquarists," serving as the secretary of the Philadelphia Aquarium Society and wielding great influence over the national hobbyist scene through his publication of aquarium books and journals.

Innes was most certainly not the only hobbyist to combine interests in fish fancying and photography, but he was, as a member of a publishing community, important in developing and spreading the use of the photographic tank throughout the hobbyist community. The publishing house of Innes and Sons was established by Innes's father; William joined in 1895. Immediately after his introduction to the aquarium hobby, Innes started publishing aquarium texts through Innes and Sons. The first publication, Herman T. Wolf's *Goldfish Breeds and Other Fishes* (1908) did not sell well, and after ten years had created a thousand dollar loss for the company. Innes eventually paid Wolf for the rights to the book and used the goldfish sections, combined with information from his own experiments and those of the New York Aquarium Society with tropical aquarium fish, in his extremely popular *Goldfish Varieties and Tropical Aquarium Fishes* (1917). Recognizing the growing interest in breeding, Innes started the first national hobbyist publication, the *Aquarium*. In 1935 Innes turned completely to the tropical hobbyist trade, publishing *Exotic Aquarium Fishes*; the book has gone through nineteen editions, updated with the help of Innes's good friend, the ichthyologist George Sprague Myers.

Innes attributed his success in publishing to his continued experimentation with aquarium photography. The earliest hobbyists identified and standardized their breeding stock via the circulation of drawings. Herman Wolf was well known for his drawing acumen, and Innes notes that his images of specimens such as the Philadelphia Goldfish and the Japanese Veiltail were "masterly" and "copied worldwide." While Innes reprinted these pen-and-ink drawings in *Tropical Aquarium Fishes*, he chose to follow his passion for photography to illustrate the newer parts of the text.[32] In addition, his hobbyist journal became known for photography, and Innes credited these visualizations with its success. According to him, "the success of our magazine [*The Aquarist*] is largely based on the quality of our photographs of aquarium fishes."[33] For the photographs in both his magazine and books, Innes developed a variety of new techniques, including a photographic tank design and a coloring process, that eventually spread to the larger marine research community.

Innes produced high-quality aquatic images for publication and became the epicenter of this craft network in the middle of the twentieth century. In an article titled "Aquarium and Fish Photography" in the *Complete Photographer*, he starts by stating that many aquarists and scientists have commonly eschewed photography as a form of documentation in both hobbyist and scientific fields because of the inherent difficulties. However, according to Innes, many of the problems of aquatic photography had been worked out. Issues of blurring because of movement and glare from scales and glass could be alleviated by combining a specialized tank system, quality camera equipment, and a set of techniques prescribed by Innes.

The first suggestion from Innes is the construction of a specialized photographic tank. There are two components to the photographic tank designed by Innes: the main aquarium and reducing compartments. The main aquarium is not where fish are routinely kept but a specially sized aquarium meant to restrict the movement of fishes while allowing them to swim freely. The dimensions of his photographing aquarium, meant to fit most medium to small fishes, are 7.5 inches high and wide (the front and back panes are 7.5 × 7.5) and 3 inches wide. This resulted in a tall, thin tank that could be moved easily for photographing inside (artificial light) or outside (sunlight). If the photographer wanted to photograph a smaller fish, reducing compartments could be added to restrict the movement of the subject.

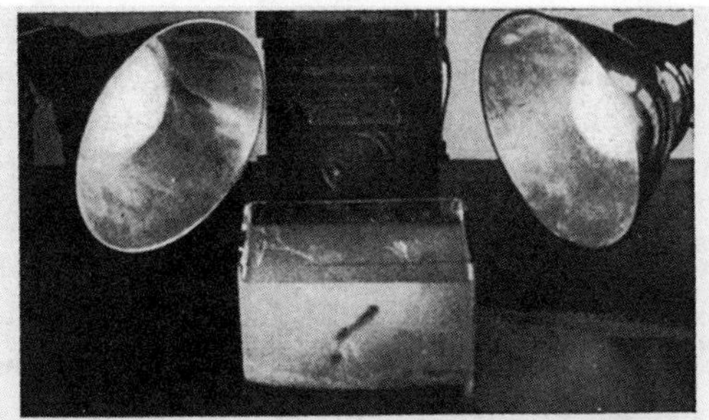

Showing the arrangement of lamps on a small tank

Taking a photograph from above

Showing the position for a background. The color or shade of the card can be varied to suit the subject

FIGURE 2.3 Several setups for photographing small tanks. Here, Innes demonstrates how his use of lighting and background shifted depending on the position of the camera. William T. Innes.

Innes gives two versions of reducing compartments. The first is a three-sided glass structure referred to as a "fish cell." When placed in the aquarium and fitted flush with the side of the tank, it formed a smaller compartment for restricting movement. The second reducing compartment consisted of building notches into the top of the main aquarium every three inches. To create a reducing compartment the photographer would fit a 7.5-inch pane of glass into the notched area, thereby reducing the overall width of the main tank.[34]

In addition to tank dimensions, Innes shared his knowledge about camera equipment and techniques. Lighting issues could be solved in a variety of ways. The first was to build the bottom and sides of the photographic tank with white porcelain. The reflection from this white surface meant that he only needed minimal lighting when working indoors (a single flash bulb instead of an arc lamp or magnesium flare). If one did not have access to flashbulbs or good lighting, he suggested working with daylight. According to Innes, the best lighting occurred when the sun was at a forty-five-degree angle; photographers should steer clear of early morning or late afternoon to minimize glare and shadows. In addition to lighting, background color could solve many problems with visuals. Innes suggested a white or gray background for fishes with opaque fins and a black background and shooting toward the light to capture the definition of clear fins. Similar to Shufeldt's umbrella adaptation, much of the equipment suggested improvisation with household objects. Innes's outdoor photography platform was an ironing board. His funnels for placing fishes in cells or tanks and his backgrounds were handmade with cardboard. And all of his pushers and cells were held in place with wooden clothespins.

When it came to color photography, Innes combined earlier forms of natural history illustration with photographic technology. According to Innes, color photography took too much light exposure and still failed to adequately capture the actual coloring of the specimens. To counteract this problem, he shot photographs of fishes in black and white and then hand painted the plates until he felt that the colors matched those of the living specimens. The resulting images were brilliantly colored and prized as both scientific and artistic objects.[35]

William Innes's photographic work made an impact in both photography and aquarium hobbyist communities as well as the academic research community. He corresponded with a number of academic researchers interested in using photography in their work. For instance, in 1930 Myron Gordon sent a group of images to Innes to critique. Gordon was a cancer researcher at Cornell and the New York Aquarium using platyfish to study the link between cancer and genetics. According to Gordon, he had "taken some autochromes of platys and hybrids but the average price is so high I cannot do it on a large scale." Gordon wanted to take color photographs of all his fish for publications but felt that he would have to use Innes's method of hand painting to get true color. Gordon closes the letter saying, "I confess that my first ideas of fish photography came from reading the chapter on fish photography of your beautiful book."[36] In addition to Gordon, Innes worked and corresponded regularly with Charles Breeder Jr. of the Museum of Natural History, New York University, and the New York Aquarium, and Henry Fowler of the Academy of Natural Sciences in Philadelphia.

Innes's working relationship with George Sprague Myers tells us much about the power of photography to facilitate conversation and knowledge production across disciplinary boundaries and about the importance of Innes as a knowledge courier from hobbyist to academic communities. Myers was the president of the American Society of Ichthyologists and Herpetologists, and he worked at Stanford for much of his career. He served as the scientific adviser to Innes for his books on tropical fishes and corresponded regularly with him. Besides being one of the most lively and downright pleasant correspondences I've ever read, these letters show cooperation between hobbyists and academics through the exchange of photographic specimens. For instance, in 1931 Innes sent a group of photographs to Myers that caused quite the reaction:

> I nearly had apoplexy when I got those pictures last night. I could scarcely take time to eat dinner before I dashed out to the Museum to see what I could do with them. Such fishes! And I don't suppose you have a pickled specimen of one of them to make sure what they are. . . . This is the most heartrending of all. What is it? Here I am a specialist in Characins and you send me a beautiful photo of one I've never seen before, alive or pickled——with no specimen. It surely is no Poecilobrycon or Nannostromus if it is as large as the photo is; all of them are tiny things. It may be a Hemiodus, or a Rhytiodus, but I can't tell. You must, by hook or crook, get me one of these, for it looks decidedly interesting. Very likely it is new. Is there any data as to locality? Probably the Amazon.[37]

From this letter, we can see that not only was Innes in touch with academic researchers but also that photographs were being circulated as proxy specimens (although they still cannot stand in for the actual physical specimen). Of particular interest here is the way that Innes not only provides the photograph but also the ability to read it—including the way that he might have sized the fish. Obviously, Sprague wishes that Innes had collected these specimens in the way of academic scientists, but the photographs allow the two men to speak with each other in a way that might be otherwise impossible because of different training and motivations.

Myers was not the only academic biologist with whom Innes collaborated. Carl Hubbs, an ichthyologist at the Scripps Oceanographic Institution, coauthored an article on a new species of blind fish with Innes. This article, titled "The First Known Blind Fish of the family *Characidae*"

FIGURE 2.4 First photographic illustration included in an academic marine taxonomic publication. William T. Innes. Hubbs and Innes, 1936.

(1936) was the first academic journal article to use a photograph exclusively as illustration of a new species. Until this time, a photograph was always accompanied by a drawing of the species, with line drawings and sometimes smaller drawings of particular features to illustrate the important morphology. The inclusion of a single photograph as illustration shows the way that photography had advanced enough for academic researchers to accept them as forms of truth claims.

Innes is rarely referenced by later scientists or photographers. The nature of craft knowledge is that it often continues when the source of that knowledge fades. The fading of craft sources is even more pronounced when that knowledge moves through a wide network of disciplines and users. Commonly, knowledge is attributed to the first person in the discipline to use the craft knowledge. In aquarist and photography circles, Innes was acknowledged into the late twentieth century. Sam Dunton, the photographer at the New York Zoological Society, said this about Innes in his 1956 book on animal photography:

> I learned fish photography the hard way. Taking pictures in the tanks of the old New York Aquarium in Battery Park in the early Thirties was my introduction to professional nature photography. I have been preceded by such pioneers as William T. Innes, Louis L. Mowbray, Sr., and Elwin R. Sanborn, to mention only a few. Actually, I have always regarded Bill Innes as the pioneer photographer of home tropical, and his magazine *The Aquarium* certainly proved a good many years ago that tropical fish enthusiasts can take excellent pictures of their aquatic pets in color or monochrome. Innes was publishing full photographs of fishes in the late twenties. The results he obtained with the old Agfa Color Plates, long before the advent of modern color emulsions, encouraged me to try my hand at the color photography of fishes in 1933.[38]

This is high praise for Innes, but it did not extend to the more academic disciplines.

Even though Innes collaborated with academic researchers, his contribution to the more formal scientific disciplines was largely forgotten by the 1960s. Another photographer, Herbert Axelrod, is occasionally credited with introducing Innes's photographic techniques to academic circles. Axelrod, born in 1927, was considered a wunderkind in the tropical fish field. At twenty-four he wrote *Tropical Fish as a Hobby* and published *The Handbook of Tropical Aquarium Fishes* only three years later. He received a PhD in biostatistics from New York University. By the late 60s he had made his name and fortune as the largest publisher of pet books worldwide, and his aquarium magazine, *Tropical Fish Hobbyist*, was billed as "THE ONLY AQUARIUM MAGAZINE IN THE WORLD ILLUSTRATED INSIDE WITH COLOR PHOTOGRAPHS!!!" In a 1965 *Sports Illustrated* interview, the journalist stated that Axelrod "regards himself as the finest photographer of tropical fish in the world." Axelrod's magazine was not the first with color photographs, nor did he take all of the photographs himself. William Innes sued Axelrod for stealing many of his color images of tropical fishes without compensation. But while Innes won that lawsuit, Axelrod's personality, his PhD, and his bombastic self-promotion as the world's best fish photographer managed to erase much of Innes's contribution to the field of marine scientific photography.[39]

Photography Enters the Academic Marine Sciences

Before the middle of the twentieth century, photography was not fully integrated into marine science. As previously noted, pioneers of aquatic photography recognized the potential of their work for academic biologists. But Shufeldt's belief in the accuracy of photography over draftsmanship and Innes's use of photographs as specimens for taxonomic analysis over distance (especially without the use of preserved specimens) did not immediately make photography and the photographic tanks useful in the laboratory. As we saw with Gordon's letter, one prohibition for the use of photography in aquatic labs was price; another was the inability to get these images published in journals. Until the middle of the twentieth century, the use of photography in aquatic sciences was sporadic, relying largely on the interest of the researcher and journals willing to publish photographs instead of pen-and-ink drawings.

After World War II the growth of photography in aquatic laboratories and fieldwork rose dramatically. There are two possible reasons for this growth. The first is the advancement of photographic technology. In aquatic science, color is often the key to subtle taxonomic and behavioral distinctions in species. While color photography existed before this period, as we saw with Innes, concerns about the objectivity of that coloring led many researchers to continue to rely on hand coloring of plates to attain what they perceived as accuracy. Post–World War II, shutter speed, film quality, and lighting apparatus finally produced color images that researchers felt accurately captured the physiological aspects of specimens. In addition, cameras became less cost prohibitive, as did film, and what was once a higher-end hobbyist interest became more common. Even the lowest-funded laboratory had access to a relatively good camera, and in many cases, the camera might save money. Film was certainly cheaper than hiring experienced draftsmen to illustrate scientific papers.[40]

As is often the case, when photography was accepted as mainstream academic practice, the ties to nonacademic groups were largely forgotten. In 1941 Lorus J. Milne placed a small note in *Science* describing a "simple, thin aquarium" built for use in photographing live fishes. The images presented are nearly identical to Innes's but contain no citations or mention of his work.[41] In fact, Innes's work has not been cited in any academic publication on aquatic photography—even when Shufeldt and Boutan are cited or gestured to as predecessors of the field. Instead, most academic biologists working on photographic tank construction post–World War II cite a single progenitor of their craft: John "Jack" Randall. Randall's one-page article "A technique for fish photography" was published in *Copeia*, the official journal for the American Society of Ichthyologists and Herpetologists, in June of 1961. The article had no images and no citations but has been cited by nearly every article on taxonomic photography since its publication. And the growth has been rapid: today, photography is integral to academic aquatic biology and especially to fish taxonomy. We will examine the rapid growth of photography in academic science by following the development of both tank design and technique between 1961 and the present.

Jack Randall

Jack Randall (1924–2020) was a world-renowned ichthyologist and fish photographer. He received his PhD from the University of Hawaii and, after a peripatetic career indicative of most marine biologists, was named senior ichthyologist at the Bishop Museum in Hawaii in 1970. He remained (emeritus) at the Bishop Museum until his death in April 2020 and still answered emails from curious historians at the age of eighty-eight. Randall grew up in California and developed an interest in fishes early in his childhood. He maintained saltwater aquariums from a young age and continued to do so throughout his adolescence and young adulthood. His interest in the shoreline of California would eventually lead him into a career as a marine taxonomist. Randall served on several survey teams to French Polynesia and specialized in the classification of coral reef fishes of the Indo-Pacific.

Randall's 1961 paper marks the introduction of photography into mainstream academic aquatic research. In the short paper, titled "A Technique for Fish Photography," Randall introduced a technique for capturing color photographs of specimens that would be taxonomically useful for the scientific community. Accuracy of color and spine count are important aspects of species identification, but too often photographers worked with subjects that rapidly lost color, had been damaged, or had shriveled, shrunken, or folded fins. Randall's technique involved transporting a recently caught specimen to the laboratory and placing it in a wax lined pan. The wax lining allowed the photographer to carefully pin the fins of the fish open and paint them with formalin. Formalin was commonly used as a preservative on collecting trips. Randall states that pinning fins and carefully painting them with formalin fixed them in an upward position and allowed the photographer time to repair any damage to the specimen that occurred during capture. After the formalin had dried, the specimen was placed into a photographic tank with a black or gray background and photographed. This technique combined the photographic tank technology with the new technique of using formalin to process a dead fish.[42]

Because Randall was introducing a new technique for preparing specimens for photographing and not focusing on tank design, he does not formally cite any previous tank designs. In fact he only mentions one other person in the article: Charles Cuttress of the National Museum of Natural History in Washington, DC. According to Randall, Cuttress developed a very interesting technique for tank lighting using various pieces of plastic sheeting to direct light into the tank from below.[43] Aside from this mention, Randall does not identify predecessors, either from the academic sciences or the hobbyist community of his youth. When I spoke with him, he did not recall a single place he had heard about this technique or anyone who had really taught him how to photograph fish. Instead, it seems to have been in the air during the period without a formal citation to pinpoint. But after Randall's publication, a flurry of techniques and technologies sought to build on his work, and photography formally entered academic taxonomy.

Shortly after Randall's note in 1961, ichthyologists began to publish updates on both the technology and methods used to capture images of aquatic organisms. These publications were largely concerned with field images. Capturing an image of a specimen while it was still living, and the possibility of returning that specimen to the wild after it had been photographed and entered into field notes, made field photography techniques especially sought after. Each publication built on the last, with Randall's 1961 piece the agreed starting point in the literature. What is interesting about this agreement is that many of the "new" tank designs and methods resemble those previously highlighted in this chapter. Tracking these publications demonstrates not only how the aquarium tank and subsequent photographic techniques became formal knowledge in marine-biological study but also how previous knowledge gets folded into this formality, erasing both individuals and whole groups of knowers.

Field Photography after Randall

While many in aquatic biology continued to rely on line drawings and illustrations for creating images in the field, some researchers in oceanography and field biology turned to photography as a way to ensure accurate color depiction. Of particular interest was developing a method to photograph specimens on boats. Because any fish caught would need to be transported back to shore, unless photographic tanks were developed for shipboard use all illustrations and photography would be done on dead and preserved specimens. The longer a specimen stayed out of water, the more it lost its coloring and other important characteristics.

FIGURE 2.5 Small internal tank and a pliable partition. David's setup used a single tank for multiple size specimens. P. M. David. United Kiosk-AG.

To get the best quality images, shipboard photographic tanks needed to be developed.

The first researcher to publish on a shipboard system was P. M. David. David was a researcher at the National Institute of Oceanography in Surrey, England. According to him, improving photography on research vessels could improve biological oceanography and link field and lab spaces by providing those on land with accurate images of marine species. David suggested anchoring the tank to prevent movement at sea and using short duration flashes to counteract vibrations from onboard machinery. Similar to Innes's design, he built a tank that was glass on four sides but with an insert inside that kept the specimen in a prescribed space that was both level with the camera lens and angled for the best lighting. The insert, made of a plane of glass in the front and a thin sheet of flexible plastic (Perspex) in the back, closely resembles Innes's glass photographic aquarium insert. David's use of plastic made it more accommodating for different size specimens; he also used a hard, clear plastic to attach the photographic tank in the larger aquarium structure.

FIGURE 2.6 Photograph from P. M. David's publication. The difference between P. M. David's illustrations and those in academic journals demonstrates the emphasis on aesthetics and artistry in David's work and the goal of the publication. P. M. David. United Kiosk-AG.

With this setup, David produced very clear images of pelagic (open ocean) species. While David was certainly aware of the scientific importance of these images, it is also evident that he recognized their beauty and the artistry involved in the marine photographic craft. His paper on this technique was published in a photography journal, not a scientific research journal. Other scientists cited his work, but he never published in aquatic research journals even though he himself was an oceanographer. It is not clear whether he trained his own students in these methods, but they have been cited by others.[44]

In 1966 T. F. Pletcher, a Canadian fisheries biologist, tested a variety of aquarium designs and photographic methods at the Vancouver aquarium and, eventually, onboard a fisheries vessel. His published method included schematics for a tank design similar to others in this chapter—

including a movable glass partition for containing the subject and the usual information about background lighting and coloring. But the most important aspect of this paper was the method of anesthetizing the subject. Shortly before photographing, Pletcher suggests injecting the fish with 2-Phenoxythenol. The anesthetic creates a tetany of the fins, basically subduing the specimen while tightening and extending the fins. However, it didn't produce this tetany in every species, and it sometimes caused muscle contractions and twitches that could interfere with photography. Pletcher suggests placing the specimen in water treated with M.S. 222 (a common anesthetic for fishes) for a minute and then injecting 2-phenoxythenol under the gill covers right before photographing for optimal results. If photographing lasts longer than five minutes, photographing the fish inside the M.S. 222–treated water is suggested.

Pletcher's conversational tone and lack of citations suggests that much of what he is imparting is craft knowledge to many in the field already; the author is just combining it and publishing it for posterity. The suggestion of using a hypodermic needle to remove air from the swim bladder of fishes that float and scraping air bubbles off the glass with squeegees before photographing are minor elements of the paper. These were usually steps discussed in personal letters or learned through tacit knowledge in laboratories. In fact, much of the conversation about anesthetizing subjects seems to be written as common knowledge, not a personal finding but instead something Pletcher heard about and passed along with other techniques.[45] The lack of citations in the paper and the lack of citations of the paper suggest that those in the field employing photography might not consider the techniques formal knowledge and therefore do not feel the need to cite them.

In 1980 Alan Emery and Richard Winterbottom combined twenty years of publications and conversations about fish photography into a single publication. Emery and Winterbottom published both schematics and methodological steps for obtaining clear images of large quantities of specimens in the field. According to the authors, on a research trip to the Chagos Archipelago in the Indian Ocean, they photographed 120 specimens a day. Emery and Winterbottom sought to show their fellow ichthyologists a foolproof way to do the same. Their paper suggested building three tanks with designs similar to Pletcher's and using Randall's formalin technique. The main advancement to the field was a set of step-by-step instructions combined with a very thorough bibliography. In fact, the authors state,

> Undoubtedly there are many aspects of fish specimen portraiture which we have not solved well, but this method worked for us under difficult

> circumstances. As there are no easily available instructions, we offer this as a starting point, and hope that readers who feel that they have improved on this basic system will communicate with us.[46]

In the short opening literature review, they call attention to publications (including Randall and Pletcher's) but also highlight Dr. Walter A. Starck II (a coral biologist and underwater photography pioneer), Gerald Allen (ichthyologist), and Herbert Axelrod (ichthyologist and hobbyist publisher, including *Tropical Fish Hobbyist*), none of whom had published anything about their methods. Highlighting unpublished knowledge producers in the text of the article instead of in the acknowledgments section is an indication of how important craft knowledge is in this field. In addition, the individuals highlighted were not from a single area of study—a coral biologist, ichthyologist, and a hobbyist publisher each contributed to the authors' knowledge. This shows us that although this period allows us to follow technology and technique through citation, we should in no way believe that the movement of knowledge and ideas was limited to publications and papers within a given disciplinary boundary.

After Emery and Winterbottom's publication, small changes to tank design and methods were published throughout the second half of the twentieth century; however, no major changes were made to the basic tank design, and the suggested methods are relatively minor changes. Many papers introduced plexiglass tanks, and others compared plexiglass to glass for fieldwork and lab work. The general consensus seems to be that plexiglass tanks are fine for short-term work but that they are easily scratched during field movements. Another aspect of glass is that it can be assembled quickly in the field by carrying the pieces flat and then gluing them with aquatic cement or epoxy that sets relatively quickly.[47] The argument over plexiglass versus glass photographic aquariums remains a matter of project and personal taste, but it is still common to use plexiglass in the laboratory and glass in the field.

Another issue was the concern about fish mortality in the field, especially in organisms that could be threatened or endangered. The desire to reduce fish mortality and induce natural behavior from fishes resulted in Rinne and Jakle's *photarium* (1981). This tank, built out of plexiglass, was attached to a system that pumped water directly from the subject's water source. The fish was placed in the system, and the current and comfort of home conditions facilitated natural behavior. Rinne and Jakle felt that this method was better than Pletcher's anesthetic method.[48]

There are many interesting aspects of these publications, but what is most intriguing is the tinkering shown in these papers. Instead of pursuing agreed-on methods for standardization, these papers are written as general guidelines even when technological schematics for tanks and measurement for anesthetics are given. For instance, each author mentions their background material. Old pieces of cardboard painted white, gray, or black, a piece of plywood wrapped in gray fabric, and even a background made of old sheets from home. To reduce glare, photographers suggest standing behind a cardboard screen painted black with a cutout for the lens; others suggest wearing dark clothes. A major focus in all these papers is not necessarily changes to get the best images, but to produce them at the lowest cost. Many authors give estimates of material costs and actively compete to produce the cheapest setup. For instance, Rinne and Jakle spent about one hundred dollars on the pump and portable battery for their tank, but the rest of the components cost them less than fifteen dollars.[49] Overall, these papers do not seek to standardize the process of fish photography but instead to communicate possibilities.

Expanding Oceans under Glass

Today, photography is an integral tool for marine science in both the laboratory and the field. Digital photography has changed some of the process; researchers can instantly review their images to assure clarity and positioning, and scanning has become a popular method of imaging.[50] But many things have not changed. Jeff Williams uses a tank setup very similar to those used by his predecessors. He assembles his tanks from plate glass onboard ship with fast drying aquarium cement. He carries a precut plate of glass to use as a positioning screen and keeps it secured with clothes pins. His backdrops are made of simple cloth draped over cardboard or plywood. In one image of his setup you can see his use of a field-assembled tank. On the bottom of the tank are notches into which he fits the glass partition to contain the specimen; that partition is held in place with clothes pins and a piece of wood. The fish are killed with rotenone and brushed with formalin before photographing. A simple, reusable plastic bin is balanced on top of a rattan chair to raise the tank to eye level. Williams's technique is reminiscent of Innes and Randall; while camera equipment and lighting continue to change, the basic setup of the photographic aquarium does not.[51]

Despite the long history of photographic tanks in both practice and publication, it remains a process largely learned through tacit and craft knowledge. When asked how and where he learned to build tanks and photograph fish, Williams was momentarily at a loss to pinpoint an exact time or place. He did not take formal classes, nor did someone necessarily "teach" him how to build a photographic tank. He said, instead, that he learned during his MA and PhD by watching others in the lab and tinkering on his own. Williams did mention Randall's work as inspiration and pointed out differences in their techniques (they use different background colors). In our email exchanges, Randall pointed to Williams as an example of great technique. Formal academic publications about the process of photographing fishes have nearly stopped. In 2013 the book *Imaging Marine Life: Macrophotography and Microscopy Approaches for Marine Biology* became one of the first published sources of information for marine academics who wish to use photography in their work. The book contains a variety of information about marine photography and gives special attention to the use of underwater cameras and photographing organisms in situ. Out of ten chapters, only one (chap. 9), shares information on the use of aquariums to capture images of marine organisms, and the majority of that chapter (over 75 percent) focuses on teaching researchers the basics of photographic equipment and terminology (aperture, exposure, flash).

The aquariums described in the chapter are simple, with a small tank balanced on two Styrofoam blocks for height and containing a black plastic sheet to corral smaller organisms into the center of the shot. Most importantly, the authors do not give a single citation to explain the source of this information. As an academic publication, it is clear that attribution is important. They do cite Haeckel and Dionysus when talking about the importance of images in natural history, but they do not cite a single technical paper on the history of aquarium photography. The authors might have learned about their craft the same way that other scientists have—through the passing of craft and tacit knowledge in the laboratory and tinkering in the field. It makes sense that they did not cite information that they did not know exists. The history of tank photography and the way that the craft has developed continues to be ephemeral, as technique is usually discussed person-to-person, and this occurs via email and discussion boards, not formal publications.[52]

Following the history of photographic tanks shows both the wide network of actors who contributed to the current methods and tools and the craft nature of the knowledge needed to work these tanks. Photography hobbyists and aquarium hobbyists first began tinkering with tanks to develop the craft knowledge needed to take clear images. But knowledge couriers helped spread photographic craft throughout the network of users. William Innes, a largely forgotten figure in the field, straddled many communities, including photographic hobbyist, aquarium hobbyist, and academic researcher. His work with leading scholars helped push photography into mainstream academic publications.

Today, photography is an important aspect of modern marine science, and it is still equally important for the aquarium hobbyist.[53] Both use photographs for truth claims. Technologies and methods developed throughout the twentieth century continue to be used today, and new methods are developed for the advancing technology of digital cameras, editing software, and extreme macro and micro lenses. The process of learning how to photograph fishes for scientific publication is the same as it is for photographing them in public or home aquariums: tinkering to find what looks good based on what others have done before you and what you have observed. While this might not sound particularly scientific to some, it is indicative in many ways of the nature of marine research. Collecting, constructing, and envisioning a marine environment terrestrially requires a combination of formal, craft, and tacit knowledge from a variety of knowledge producers. While many of these producers have been erased over time and their knowledge elided into the general techniques and tools of the trade, the influence of the wide range of tinkerers on the development of this craft cannot be denied.

In addition, the images produced are a combination of personal preference and craft techniques to attain individual and professional photographic goals (scientific, hobbyist, artistic). The earliest tank photographs equated to dioramas. These staged scenes contained a variety of fish and invertebrates in an environment strategically populated by rocks and plants. These scenes relied heavily on underwater imagination instead of observation.[54] But photographic tank scenes were not only built around the marine imaginations of those building them but around the requirements for the attainment of clear images. Rocks and plants had to be clean and cleverly chosen to add dimension while maintaining water clarity. Glass partitions could prevent fish from hiding in the backgrounds, but they could not hide particulates in the water. In addition, some organisms proved easier to photograph; octopuses and other majestic but often sedentary creatures made appearances in photographs more often than delicate, fast-moving fishes. While taxonomic images have a tendency to be more standard, even those produced for academic publication contain choices that can affect the viewers' ideas of the submarine world.

In a 2020 paper describing a new species of fairy basslet from New Caledonia, an image of a preserved paratype of the fish located at the Bishop Museum in Honolulu is included. The photograph, taken by Jack Randall, demonstrates the effects of his background choices on visual reception of the photograph.[55] Randall's black background is good for defining the edges of a specimen for taxonomic identification, but it also gives the visual sense of ocean depth. Juxtaposing this image with the first in this chapter, we can see that a small detail, the color of the background, makes a large difference in the photograph. While both convey

FIGURE 2.7 Image taken by Jack Randall of a new species of fish collected in New Caledonia. Randall preferred dark backgrounds to make it easier for taxonomists to see coloration and the number of spines in the fins. Jack Randall. Used with permission of the Bernice Pauahi Bishop Museum, Honolulu, Hawaii.

the morphological details required of the academic community, the individual choices of the photographers potentially change the way the viewer reads the image.

Photographic tanks are part of the way that aquarists have created knowledge about the ocean. These tanks allow the user to visualize submarine worlds for extended periods and to transform those ephemeral images into permanently circulating objects. It is through the photographic capture and circulation of these images that a wide community of researchers have come to share a vision of the submarine world and to know more about the creatures that inhabit these previously hidden spaces. The artistic nature of these images and the individual choices made by photographers does not negate the usefulness of photographs to taxonomic identification in marine science. But understanding their creation also helps us understand how science and art are continuously intertwined in the use of tanks. As we continue to explore more tank varieties, we will see how oceans under glass expand through these images and construct our perception of the submarine world through that expansion. In the next chapter, we see the development of a tank meant to replicate an essential element of the ocean: its physical movement.

3

Kreisel Tanks

Crafting Movement under Glass

In 2011 Alex Andon launched a Kickstarter campaign to raise funds for his company Jellyfish Art (JFA). Andon, an entrepreneur with an undergraduate biology degree, hoped to develop the first commercially available kreisel tank. Kreisel tanks (from the German word for "spinning top" or "rotation") have been used for over a century to maintain jellyfish in laboratory and home settings before Andon's Kickstarter campaign. But the design was not patented, nor was it commercially available to true amateurs uninitiated into the aquarium hobby. Jellyfish Art (JFA) sought to change this. Andon advertised "the first affordable aquarium designed specifically for jellyfish" and promised that it was "as easy to maintain as a regular fish tank." The Kickstarter goal was originally set at $3,000 but quickly raised $162,917 with 515 backers set to receive the first iteration of the acrylic tank, moon jellies (*Aurelia aurita*), food, and saltwater mix.[1]

When these backers began receiving their tanks, Andon's promise of easy maintenance began to unravel. One user stated, "I want a refund. You misrepresented the ease of taking care of jellyfish. I took my tank down after my jellies died after less than two weeks being in the tank. I spoke to several professionals in the aquarium field and learned that jellies are very difficult to keep alive in captivity." Many users stated that their jellyfish lived less than forty-eight hours; they were either sucked into the water intake valve or caught on the substrate lining the bottom of the tank.[2] Others created trouble-shooting techniques and shared them on special social networking pages. On the Facebook page for JFA, users shared tips for fixing faulty pumps (a nine dollar pump from the local pet store), noisiness (pinching the air pump with a chip clip), water level problems, feeding issues, and many more. These fixes took time to figure out and required an enormous amount of attention to a system that was promised as hands off. In fact many of the commenters on the Facebook and Kickstarter web pages quickly transitioned into serious leisure hobbyists—building their own tanks, rearing their own brine shrimp (instead of purchasing it from JFA), and ordering and rearing other species of jellies from established hobbyist companies.[3]

Despite the troubles with JFA's initial roll out, the mass-produced kreisel tank market quickly took off. Andon's initial tank design was quickly followed by updates from JFA as well as tanks offered by new companies including Eon and Cubic Aquarium. And JFA came back into the limelight in 2015 when Vanilla Ice featured their tank on his building show. While many of the systems start at only $150, the higher-end (and higher-functioning) Eon and Cubic systems start at a much higher price. The changes to systems and the higher price tags do not mean that the user needs less information about their system. Eon users troubleshoot and share information via Facebook pages and other resources such as MoonJellyfishblog.com or look for information and share tips and tricks on the Jellyfish Keepers board. All these companies offer a slightly different tank design, but their users run into similar problems: "sad" jellies that are either bruised, unable or unwilling to feed, or just generally unhealthy.[4] Kreisel tank entrepreneurs, and the amateur hobbyists who patronize their businesses, are learning the hard way that building a kreisel tank is only the first half of keeping jellyfish in captivity. The methods for maintaining these organisms in captivity are equally important as (if not more important than) the enclosure in which they are kept.

A variety of globally dispersed tank users developed the craft required to pair technology with techniques over the course of the last century. The history of these tank systems, and the craft knowledge required to make those tanks function, began in the early twentieth century in physiology and morphology laboratories with researchers interested in studying nerve function creating tanks to extend the life of jellyfish in captivity. But it was only through the painstaking work of determining individual diets and behaviors of the jellyfish that the tanks functioned. While interest in working with jellyfish in neurophysiology diminished, public aquarists began maintaining jellyfish for display. Aquarists on multiple continents perfected the tank and methods to maintain healthy jellies for exhibition. Finally, oceanographic researchers rekindled interests in the physiological and behavioral mechanisms of gelatinous zooplankton and, in conjunction, the kreisel tank. Today, public aquariums and academic biologists continue to push the limits of captive care of jellies, building enclosures for deep-sea comb jellies and other gelatinous zooplankton with highly specialized habitat requirements.[5]

The development of the kreisel tank shows how users combined field research and laboratory tinkering to build oceans under glass capable of maintaining a complex organismal life cycle. Many jellyfish[6] have a life cycle that includes a sedentary bottom-dwelling period (polyps) and swimming or floating in the open-water column (ephyra and medusa). This varied life cycle required the development of technology that could mimic multiple types of water movement and circulation. Each species of jellyfish requires a different form of circulation. In addition, each part of the jellyfish life cycle requires very specific food sources for survival. Some of the larger jellyfish have a diet so varied that they can need upward of five different food sources throughout their life, from the smallest plankton and brine shrimp to other jellyfish and whole finfish. Building a technological system that can simulate ocean circulation combined with husbandry techniques that shift drastically with the organism's development required feedback between field observation and laboratory tinkering. This loop, from field to laboratory and back, eventually resulted in successful husbandry and increased knowledge about the life cycle, feeding habits, and behavior of jellyfish more broadly. Today, oceans under glass are full of many species of jellyfish, all existing in tanks that simulate the natural environment we have come to better understand based on tank work.

In addition to following the feedback of tank design and the development of knowledge about jellyfish and their essential needs, tracing the history of kreisel tanks exposes several important aspects of the way that information travels in the marine science network. Experimental physiologists and morphologists developed the earliest kreisel tanks during what might be termed the "golden age" of experimental marine stations, from 1890 to 1930. But the development of both the tank and techniques refining it for a variety of species quickly moved into numerous academic research and public aquarium communities with a wide geographical spread. After the initial development of both tank technology and feeding techniques at the Plymouth Laboratory in England, few groups worked simultaneously to advance both the technology and techniques; instead, aquarists often focused on one aspect or the other. Updates for tank technology took place at a German marine laboratory, while feeding techniques and husbandry were advanced in a Japanese public aquarium and a medical laboratory in Arkansas. These groups did not share information, and their publications show that they did not work from similar sources; instead, they all tinkered with the same systems to develop new tanks and methods. It took almost half a century for the technologies and techniques developed in various locations to be combined again at a public aquarium. The path that knowledge follows through the marine science network can tell us much about the structure of institutions and disciplines and how they function within this network.

In many cases, the intersection of many networks of marine science lies in the public aquarium. Much of this can be explained by temporal constraints on experimental systems. Scientific studies in laboratories or onboard ships are not particularly long lived. In fact, apparatuses used for experiments are often built de novo out of found objects by a tinkerer or engineer and broken down as quickly once the experiment is concluded. It is not rare to find as many duct-taped tanks and patched systems in marine laboratories as expensive pieces of specially built equipment.

The first jellyfish system, built in 1898, seemingly worked well, but data about its function ceased after the first few weeks because people had to leave the laboratory for the academic school year. That system was built out of available glassware and containers scrounged from the kitchen of the marine laboratory. But this is not the case at public aquariums. These institutions build permanent displays, and their work requires beauty and precision lest the public witness not an underwater marvel but a cauldron of death. Advancing aquarium craft for better exhibitions is one of the main goals of aquariums. When one institution has success with a new tank or organism, they publish widely, share information at conferences, and in the case of jellyfish, ship specimens throughout the network. In addition, many public aquarists straddle different disciplinary and hobbyist communities. Most public aquarists develop their husbandry skills or interest in the field through hobbyist or academic communities. This makes them perfect knowledge couriers because they often have links to other communities through which they share knowledge. The combination of semipermanence and open communication makes public aquariums an important node in the marine biology network. These spaces are innovators of technology, refiners of technique, and repositories that both store and disseminate craft.[7]

Kreisel tanks offer an important look at the dual nature of labor and knowledge attribution in the marine community. When I started researching kreisel tanks, I found an almost unbroken line of engineering attribution in citations. While some individuals used few or no citations, it was not difficult to trace the line of kreisel tank designs from the earliest in 1898 to Andon's Jellyfish Art patent in 2011. Although Andon's was the first kreisel patent, it was clear that those individuals interested in making changes to the system were generally aware of previous iterations to the technological schematics of other tanks. This turned out not to be the case when tracing the development of husbandry techniques, including feeding, breeding, lighting schemes, water chemistry, and other important aspects needed to care for an organism that goes through so many distinct life cycles. In fact, it is rare to find attribution for this knowledge in citations, and it is especially rare to find it over a long period of time—even when the information remains essentially the same.

This discrepancy can be attributed to several aspects of working in the marine community. The first is the localized nature of some of this knowledge. If you have ever had an aquarium, you might have been given advice about your tank that worked—or not. The truth is that most experts on keeping aquatic organisms in captivity will admit that balancing a tank requires a complex web of knowledge about the organisms, the tank system, and your specific locality. For instance, keeping a tank of jellies alive in Florida is very different than in Colorado, let alone in Berlin, Japan, or Hawaii. Aquatic organisms react to a wide array of the components of a system, including the base you use to mix your seawater (or if you use natural seawater), access to sunlight, what filtration system you use, and what type of organisms you have in the tank. Any shift in this set of variables can mean that your system doesn't react "correctly" to advice manuals. As marine aquariums spread, the variability of systems expanded rapidly. Every system requires localized tinkering, or what many aquarists call a "blue thumb." The lack of citations of husbandry techniques in formal literature may be an acknowledgment of the highly localized nature of this knowledge and the tinkering needed to transfer it to individual systems. If you worked from a general idea but changed that to suit your system, did you use another's work or did you modify to the point of it becoming something different entirely?

Another explanation for the undervalued nature of husbandry knowledge is how it is created and who produces it. Certain types of knowledge production and the tasks involved in performing that knowledge have traditionally been valued above others. For instance, engineering and tinkering with mechanical or technical components has traditionally been considered a highly valued form of knowledge production.[8] However, many other aspects of husbandry, including day-to-day feeding, monitoring health, and essential forms of maintenance, including cleaning, have traditionally fallen to lower-level technicians and are seen as either routine or not a form of knowledge production at all. The care of animals is considered caring or body labor (depending on the work being done), and in a laboratory setting it is often devalued.[9] Historians and sociologists highlight the gendered nature of these distinctions: technical knowledge has traditionally been seen as male while caring and body knowledge is female.[10] While both men and women perform husbandry work, theorists suggest that linking body and caring work with female—and therefore unscientific, localized, and ephemeral—labor devalues that work. Caring for animals, and producing knowledge about their care, has traditionally been devalued as a lesser form of knowledge. The lack of acknowledgment of this work, both in the technical literature and in other husbandry texts, might stem from the idea that this is not, technically, important scientific knowledge but instead a lesser form of understanding that does not need to be cited formally.

This chapter follows the history of the development of kreisel tanks by separating the components of technology and technique. There are, of course, many tinkerers who cross this convenient binary. But they are exceptions, and the rule usually holds that technological tinkerers and organismal/environmental tinkerers are often two separate groups with distinct, and sometimes gendered, roles in the aquarium community. Distinguishing these ways of craft knowledge allows us to trace the movement of that knowledge separately with the benefit of seeing where and how these groups meet to build circulation and movement into oceans under glass.

The First Era of Keeping Jellyfish in Captivity

Jellyfish (medusae) are invertebrates with several distinct life cycles. The polyp is a fixed organism that almost resembles a plant; it attaches to a hard surface and can stay in this form for extended periods. When ready, the polyp replicates by budding and producing an ephyra. The ephyra resembles a tiny jellyfish and swims freely in the water column, growing into a full-grown medusa. Full-grown medusae are sexually dimorphic—there are males and females—and eventually they will produce sperm and eggs. When fertilized, these sex cells produce a planula that swims around until it finds a suitable resting place, where it settles to become a polyp, and the life cycle is complete.[11]

The jellyfish is a relatively simple organism; it lacks eyes and any noticeable anthropomorphic features. Most jellyfish have rudimentary digestive systems. Food enters the jellyfish through the mouth and is trapped in a central gastrovascular cavity, where it is broken down and absorbed into the larger system. The waste is then evacuated out of the same "mouth." But this relatively simple organism also contains a particularly interesting nervous system. Bordering the bell of jellies are organs called otoliths that help them orient toward gravity. These structures are similar to the inner ear of humans. In addition, the bell of the jellyfish is covered by a fine mesh of nerves that function similarly to human nervous systems in that the nerves carry electrical impulses sent from any part of the bell. In humans, the nervous system contains very clear pathways (the nervous system), and damage to a given pathway can mean diminished nerve function or paralysis. The jellyfish system operates differently; if one part of the tissue of the jellyfish is excised, impulses can travel over the network in almost any other direction. Instead of following a set of channels for traveling, impulses in jellyfish travel over the fine-mesh (commonly likened to cheese cloth) structure of a nervous system. In addition to this fine-mesh system, many jellyfish can regrow excised body parts, and the neural network is intact in these regrown pieces.[12]

In the early twentieth century, jellyfish became a focus of experimental biologists interested in the evolution of life on Earth. Researchers saw jellyfishes as a link between plant and animal because of their specialized reproduction and a link between humans and organisms lower on the evolutionary scale because they contain a rudimentary nervous system. These experimentalists flocked to seaside laboratories to work with fresh invertebrates in the hopes of uncovering the origin of the nervous system in the animal kingdom. The prospect of learning how to regrow nerves and restore nervous tissue to those with injuries resulted in experimentalists working not just to bring jellyfish into the laboratory but also to learn to maintain specific specimens for extended periods under great duress. The earliest jellyfish experimenters reported similar problems as Jellyfish Art users: the jellies did fine for a short period of time but they eventually seemed to get tired and sink to the bottom of the holding tank. Many developed white or brown lesions on their skin and began to shed copious amounts of jelly, clouding the water. No jellyfish lasted more than a few days in a tank, and none thrived.[13]

This section looks at the earliest efforts to keep jellyfish in captivity for extended periods to study the physiology and behavior of these ubiquitous, but delicate, organisms.

EARLY TECHNOLOGY

In volume 5 of the *Journal of the Marine Biological Association of the United Kingdom* (1899), Edward T. Browne introduced his "plunger jar." Browne, a researcher hoping to study jellyfish physiology in the laboratory, noted that keeping jellyfish in captivity was a frustrating endeavor, and he sought to alleviate this frustration with a technological fix. After the death of many jellyfish specimens in the laboratory, Browne concluded that the major difference between the captive and the natural environment was the constant tidal movements in the ocean that bore the jellyfish aloft to the surface. He stated,

> When I have been watching medusae at the surface of the sea, I have noticed that they simply float along with the tide without often pulsating

> the umbrella. In my bell-jars the water was perfectly motionless, so that a medusa had to pulsate its umbrella in order to keep afloat, and as soon as the pulsations stopped it began to sink.[14]

Browne worked with objects on hand in the laboratory space, and he consulted Edgar Johnson Allen, the director of the laboratory, to create an automatic system that mimicked marine motion. His contribution to the process of keeping healthy medusae in captivity was the "plunger jar," a fairly simple apparatus consisting of a large ten-gallon bell jar equipped with a glass plate raised and lowered by a simplistic pulley system to create a constant wave movement within the jar. The mechanism comprised a glass plate, two rods, a wooden beam, and a small bucket with siphon and water supply. The construction is described as a large circular glass plate with a hole in the center, a rod through that center hole with the glass plate attached (and slightly tilted) on one end, and the other end attached to a wooden beam. The wooden beam has a pivot in the middle; one end is attached to the rod and glass plate and the other end has a small bucket made of tin attached. A rubber tube conveys water into the tin bucket; when the bucket fills it empties itself: through the motion of filling and emptying, the bucket raises and lowers the wooden beam, and therefore the glass plate, creating a constant motion within the bell jar. Edgar Johnson Allen said in response to the successful creation of this automatic plunger jar that he "was not a little pleased to have produced an efficient piece of apparatus from just 'a treacle tin and a stick.'"[15]

Browne's simple system was extremely effective. The first plunger jar was started in the Plymouth laboratory on September 4, 1899, and the laboratory operated a set of the jars into the 1940s.[16] Browne reports that the species *Obelia* lived "very well" for about ten days and then began to die off. This was a vast improvement; *Obelia* usually lived—or, more accurately, died slowly—after less than twenty-four hours in captivity. The plunger jar increased *Phialidium* survival time from three days to six weeks. In addition to boosting the time a specimen could survive in captivity, the plunger system kept specimens healthy and thriving. Browne reported that many grew new tentacles. The jar's water was not changed, but water was added when evaporation occurred, and fresh copepods were added as a food source. Browne stated that "these experiments I think show that it is possible to keep medusae alive in confinement for several weeks without any change of water, and that they increase in size and develop more tentacles."[17] The plunger jar advanced the ability to maintain and rear medusae, and other invertebrates, in the laboratory, but it was far from a perfect device, and others continued to search for new avenues for maintaining jellies.

Browne himself was not satisfied that the plunger jar was the ideal system for all jellyfish species. Some species lived longer than others, prompting Browne to wonder whether a "slow revolving current" would suit some species better. He suggested adding a propeller in the jar to achieve this effect. Others suggested adding a filter so that the larvae could be fed continuously but the water purity could be maintained.[18] Browne's initial plunger jar was created with a bell jar that had a significant amount of algae and other specimens in the water already. He does not say why he did this, but it is possible he was trying to replicate the algae environment of the open ocean in hopes that it would be a food source for the jellyfish. He eventually experimented with a sanitized bell jar and specific food sources, but he was not necessarily concerned with water purity. Browne's plunger jar system was not intended as a complete system for *every* jellyfish species. It was well suited for medusae and other jellyfish with limited mobility that rely on currents for movement, especially the genera *Aurelia* and *Rhizostoma*. Other species required less water movement—but closer attention to other factors such as availability of food—for survival in captivity.

Browne's colleagues at the Plymouth laboratory sought to improve on the system. Their largest gripe: the plunger jar took up too must space. In 1938 W. J. Rees and F. S. Russell published a paper describing their methods of rearing a variety of local jellyfish species from planulae to adult. According to the research duo, the hardest part was providing both food and water agitation without taking up all the space in the laboratory. To do this, they took Browne's original design and downsized it. Instead of a huge tank of water with a "plunger," Rees and Russell set up a "rocking beaker" system. A simple pulley system connected a wooden batten that pivoted slightly to move metal extensions attached to glass slides in each beaker. The glass slides moved slightly when the wooden batten was shifted back and forth, creating a rocking motion that agitated the water in each beaker. This system allowed Rees and Russell to keep a wide variety of medusae at different levels of development in the same system.

The earliest publications describing these tanks did not contain much information on how to maintain jellyfish for the long term. Rees and Russell did mention that they fed their polyps and medusae but were not specific about time or information on feeding. They introduced new seawater intermittently to "revive" lackluster or struggling polyps and, with the seawater, diatoms and plankton. Browne and Rees and Russell were mainly interested in building a system that could provide enough water movement to maintain jellies in a good enough condition for experimentation. Because the laboratory existed relatively close to the water from which the jellies were collected and new jellyfish could be collected for

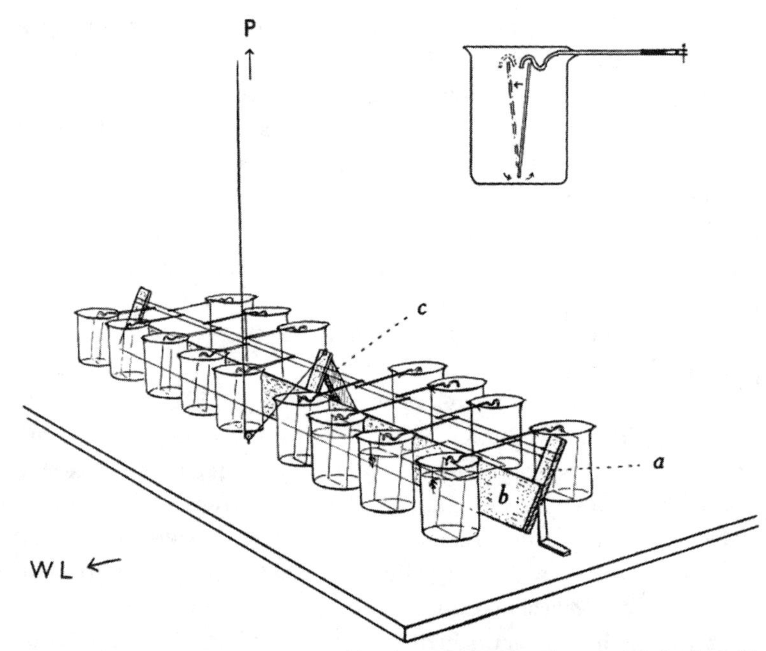

Fig. 12. Diagram showing arrangement of beakers with rocking plates for keeping hydroids. For full description see text. *a*, wooden upright; *b*, horizontal wooden batten; *c*, stopper; *P*, to plunger jar wire; *WL*, direction of window light. In the top right-hand corner is shown an enlarged drawing of a beaker and method of attachment of rubber tube to wire. (Del. F.S.R.)

FIGURE 3.1 Drawing of Rees and Russell's rocking beakers. The combination of the plunger jar and rocking beakers at the Plymouth Marine Station made it the epicenter of jellyfish husbandry in the first half of the twentieth century. W. J. Rees and F. S. Russell, "On Rearing the Hydroids of Certain Medusae, with an Account of the Methods Used," *Journal of the Marine Biological Association of the United Kingdom* 22, no. 1 (1937): 79.

laboratory use as needed, they did not worry about maintaining specimens for too long and paid little attention to their development beyond the few days or weeks required for experiments to be performed.[19] However, other people working with jellies wished to maintain them for longer to study their life cycle and behaviors. These aquarists were left to develop a better understanding of the conditions, including feeding, required to keep these organisms in long-term captivity.[20]

EARLY TECHNIQUE

Many jellyfish species could survive without constantly moving water but required a very specific diet, and determining this diet was particularly difficult for investigators. Edward Browne's success at maintaining medusae with the plunger jar was achieved without concern for the specialized diet of the specimens. Browne states that he included copepods (small crustaceans) in the original jar and replaced them as they died or were consumed, but he did not include details about how often he introduced food nor about the behavior of the jellyfish when he did. He made it clear that the plunger jar method alone would not maintain specimens indefinitely; other variables had to be attended to. Browne reported that many jellyfish varieties "died suddenly" after extended periods in captivity. Unfortunately, he did not record their condition at death (loss of weight, appearance of the pathological white spots, or just exhaustion), but it is safe to assume that many died from starvation.

Two years after the publication of Browne's plunger-jar paper, Maude Delap, a naturalist and associate of Browne's living on Valencia Island in County Kerry, Ireland, published the seminal work "Notes on the Rearing of *Chrysaora isosceles* in an Aquarium," in the *Irish Naturalist*. Delap's paper, which is still cited as a source for information on rearing jellyfish in captivity, describes her process of rearing a complete jellyfish life cycle in her home aquarium. In June 1899 Delap found a *Chrysaora isosceles* (compass jellyfish) on the shore of Valencia Harbor. She took the jellyfish home and placed it in an aquarium for future study before preservation; when she looked in the aquarium the next day, she saw small swimming forms, which she believed to be fertilized planulae. After two days, these forms had attached themselves to the side of the jar, and tentacles began to develop, signaling the beginning of the polyp stage. Delap moved several planulae to additional jars and kept the polyps throughout the winter months. By April 1900 ephyrae budded from the polyps. By May they attained a mature form and developed distinctive brown markings radiating from the center of their umbrellas (the reason for their common name). By June the mature forms required larger vessels because of size. In July the jellyfish began to struggle, and by August, they were so diminished in vigor Delap narcotized the specimens for immediate preservation. She believed that the deterioration of the specimens was due to starvation, and in fact most of her paper focused on the food provided each form of the medusae.[21]

Delap experimented with multiple food sources in each life cycle of the medusae. Her article assiduously records the food sources for her specimens, including those sources that were rejected by the developing medusae. During the polyp stage Delap stated that she initially kept the polyps supplied with copepods, "but the *Scyphistomae* [polyps], I found, preferred to feed upon small medusa, such as *Sarsia*, and little ctenophores—*Pleurobrachia*."[22] Keeping the growing ephyra and mature jellyfish supplied with food proved difficult in the later summer because of stormy weather and warm water conditions. As the supply of young medusae, especially *Sarsia,* declined, so did the health of the captive jellyfish. The death of the medusae from starvation prompted Delap to state definitively that "the chief trouble connected with rearing this medusae was to obtain a sufficient supply of food; its appetite was enormous."[23] During the mature life cycle of the jelly, Delap reports that specimens were consuming two dozen medusae and ctenophores a day. The paper includes a helpful list of what food the specimens preferred, those that it would tolerate, and the few sources that they would not ever eat.

> It had a great liking for small Anthomedusae and Leptomedusae, such as *Corymorpha*, Margelis, *Sarsia*, *Amphinema*, *Phialidium*, *Laodice*, *Euchilota*, &c.; also for the siphonophore *Agalmopsis*, and the ctenophores *Pleurobrachia* and *Bolina*. It had no objection to *Tonzoperis* and *Sagitta*. There were, however, two animals it would not touch, even after a few days' starvation—the anthomedusa *Tiara pileaa,* and the ctenophore *Beroe ovata.*[24]

In addition, Delap tried feeding the mature medusae fishes, but they only grasped the fish with their tentacles without consuming them. Eventually they let the fish go without harming them. Delap's success at rearing captive *medusae* did not stop at compass jellyfish. In the succeeding six years she published accounts of rearing *Aurelia aurita* (moon jelly), *Pelagia perla* (mauve stinger jelly) and *Cyanea lamarcki* (bluefire jelly), providing in-depth details of food sources, life cycles, and water temperatures in each subsequent publication.[25]

Delap's work was widely read and influenced other investigators interested in extending the life of captive jellyfish. Mary Lebour, a colleague of E. T. Browne's at Plymouth, combined Delap's findings on food sources with Browne's plunger jar to ascertain whether certain species actually did consume fishes as well as other *medusae* and copepods. To test their habits, she kept mature jellyfish in plunger jars at the Plymouth laboratory; because of the volume of food consumed, only one jellyfish could be maintained in each jar at a time. Lebour found that many jellyfish do eat fish, including the species *Aurelia*, *Phialidium*, and *Obelia*. She found *medusae* in general to be "miscellaneous feeders," meaning that they have a varied diet, and that they will eat many things but there is "generally some food more frequently taken than the rest," probably because of the abundance of that food source in the natural environment.[26] Similar to Delap, Lebour reported that *medusae* are voracious feeders. Lebour noted that one jellyfish consumed sixteen small fishes in the course of a half hour. Lebour's specimens survived longer and were much healthier

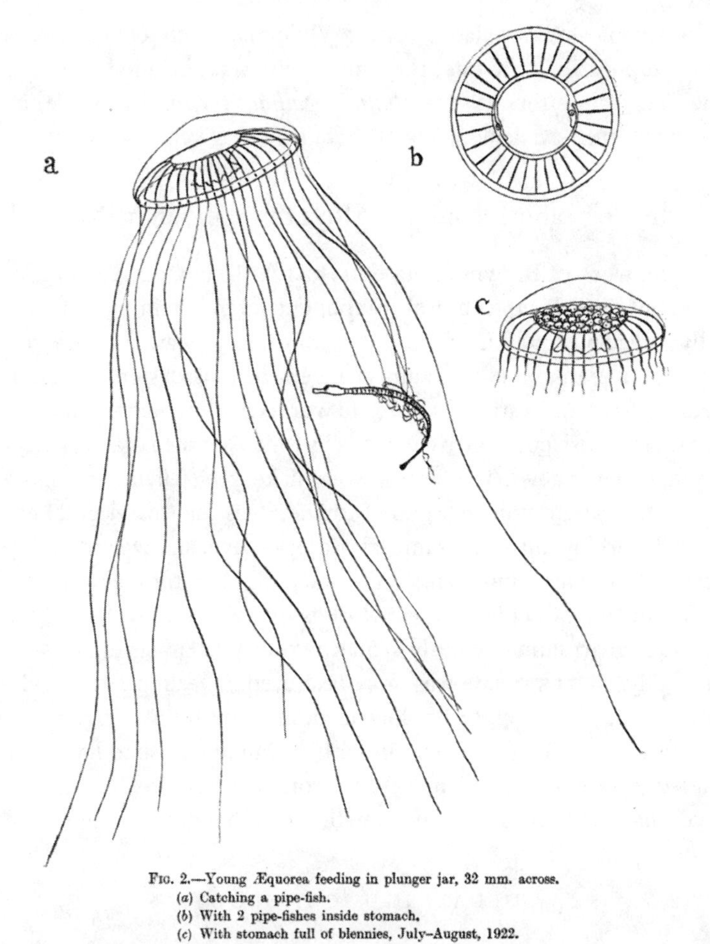

FIGURE 3.2 Drawing from Lebour's second paper on the feeding habits of jellyfish. This image shows the capture, feeding behavior, and digestion of a pipefish by a young medusae. Marie V. Lebour, "The Food of Plankton Organisms. II." *Journal of the Marine Biological Association of the United Kingdom* 13, no. 1 (1923): 76.

throughout their life cycles than Browne's, suggesting that a combination of water movement and proper feeding could effectively rear and maintain certain species of jellyfish within the laboratory for extended periods.

Although investigators worked out the process of rearing and maintaining medusae in the laboratory, few if any sought to maintain specimens in long-term captivity away from the seashore. The need for specialized vessels and live food sources meant that many investigators continued to utilize live specimens caught either at the beginning of the day or season at marine laboratories. While many different species were kept temporarily, only one, the moon jelly, was included in Frank E. Lutz's 1937 laboratory manual *Culture Methods for Invertebrate Animals* as a consistently available species capable of being kept in captivity.[27]

Technology and Techniques Move throughout the Network

After the work of Browne, Rees and Russell, Delap, and Lebour, both the technological and technical components of maintaining jellies eventually spread beyond England and academic marine stations. Researchers working on plankton in Germany changed the tank designs to suit their needs. Other groups around the world worked to lengthen the amount of time that jellyfish could thrive in tanks by tinkering with diet or other environmental variables. The earliest work on keeping jellyfish in tanks was done within a tight-knit network—Browne, Rees and Russel, and Lebour were all working out of the same laboratory space, and Delap was an associate of Browne's consistently in contact with him throughout his jelly work. This type of collaborative work is one way that technology and technique grow simultaneously in marine research. However, the second era of jellyfish tank design was more fractured, revealing a geographical disconnection between technological and technical tinkering. This section shows that many advances in tank design and husbandry occur in isolated sections of the marine network, often in separate disciplines and spaces that have little or no contact with each other during their research.

UPDATING TECHNOLOGY

While Browne's original plunger design did keep the jellies afloat, he realized there seemed to be something missing from the water movement. At the end of his work, he suggested including both the plate glass suspended above the tank to mimic surface movement and a spinning piece inside the tank (like a propeller) to agitate the water internally. Browne never composed a new technical design, but others working with zooplankton and pelagic animals at marine laboratories began tinkering with his original plunger jar.

The next technological development in jellyfish tanks occurred almost forty years after Rees and Russel at the Biologische Anstalt Helgoland, a marine station on the Helgoland archipelago in the North Sea. Similar to most international marine station research in the early twentieth century, researchers at Helgoland experimented with a wide array of living organisms to explore many biological questions. Marine laboratories contained large tanks into which collectors dumped their daily haul. These general

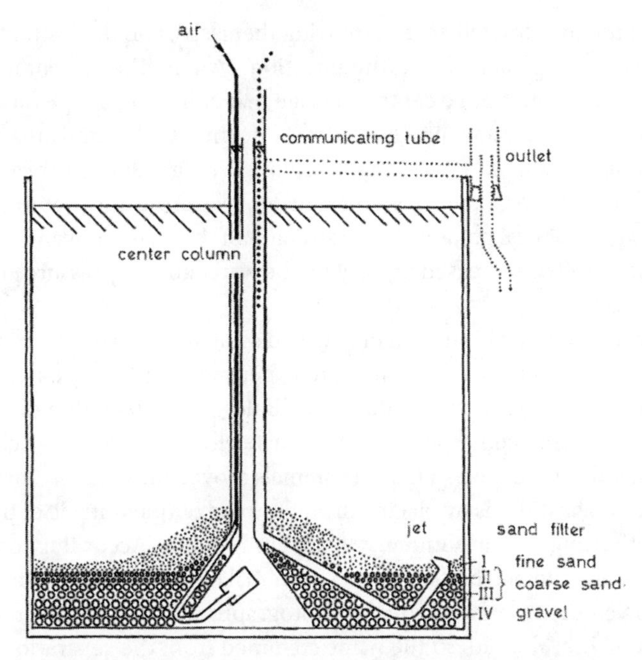

Fig. 1: Planktonkreisel. In the round jar, water rotation is caused by the air bubble water pump, which pumps the water from the bottom funnel to a higher level in the center column. (After GREVE 1968, modified)

FIGURE 3.3 Diagram of Greve's original planktonkreisel. Note the sand and gravel filter at the bottom. This aspect of the tank was also developed at the Helgoland Marine Station. Wulf Greve, "The 'Planktonkreisel': A New Device for Culturing Zooplankton," *Marine Biology* 1, no. 3 (1968): 201. Used with permission of Springer Journals.

tanks were meant to maintain these organisms in good condition until they could be observed and experimented on. They also often served as a sort of public aquarium for visitors to the laboratories. Aquarists tinkering with these general tanks at Helgoland resulted in the development of sand filters for larger tanks; the filters cleared waste out of the water and trapped detritus in the sand layers while allowing a softer substrate where animals could settle or burrow. In 1968 Wulf Greve combined this filter system with a new tank design meant to sustain gelatinous zooplankton, a group into which jellyfish fall, in the laboratory for extended periods. Like Browne's plunger jar, the tank was circular, and the main function of the tank was to maintain water movement. However, that is where the similarities end.

Greve's tank—referred to by researchers at Helgoland as the *Planktonkreisel*, translated from the German as "plankton top" (the children's toy)—contained a fine sand filter on the bottom of the tank to both filter the water and stop jellies from hurting themselves on the bottom of the tank or getting sucked into the filtration system. The system of water movement came from a central passage that contained a pipe for air and water; both water and air were pumped down into the sand area and inserted into the tank from a small output valve right above the sand line. This combination created a constant movement of the water column around the central column.[28] The strength of the water movement could be decreased or increased by varying the water and air pressure entering the system.

Greve modified his design in 1970 and again in 1975. His first modifications came about as he continued to experiment with zooplankton. For instance, his *Doppelküvette* (double cylinder) was a tank designed to do experiments on feeding and metabolism in jellyfish. First, Greve changed the airflow of the tank so that the animals moved horizontally instead of vertically; he did this by placing the inflow and output outside of the tank and offsetting the air pump inside the sand slightly. According to Greve, this change allowed better viewing of animal behavior, and consequently it allowed the researcher to take photographs of that behavior for comparison. The "double" in the name stemmed from the separation of the tank into two modules. These two tanks, side by side, had the same water and air filter system (all the water and air mixed and was filtered in the sand bottom before entering the tanks via the two separate air pumps on each side). This setup allowed Greve to keep the control group in one side of the tank and the experimental group in the other; the partition assured that the animals were kept apart but that they were in the exact same water conditions for accuracy in comparison.[29]

In the same paper, Greve also introduced another modification to his original system: a *Phytoplanktonkreisel*. Greve's overall research agenda was to understand the ecological significance of gelatinous zooplankton in the North Sea. In particular, he wanted to understand what they ate, their metabolism, and their reproductive cycles. According to Greve, while the doppelküvette and traditional planktonkreisel were useful for monitoring feeding and reproductive behaviors, the sand filter disallowed a measurement of the smallest seston—the living and nonliving matter floating in the water column—brought into the tank with the zooplankton. To capture and measure this material, Greve modified the tank bottom, replacing the sand filter, where the seston usually settled, with a perforated PVC bottom. The perforations were too small to allow the larger jellyfish and zooplankton to pass through but did permit the smallest organic materials to be collected for study.

Greve's final change to his original design came in 1975. His newest design, which he named the *Meteor Planktonküvette*, was meant to be used on board ship. To make this possible, Greve fit the top of the tank with a chimney attachment to make it possible to control the water level without it splashing out in turbulent conditions. For this instantiation of the tank, Greve kept the horizontal water flow so that the user could walk around the tank in every direction to follow the organism as it floated in the water column.[30]

Greve's designs are important in many ways. First, following his tank development reveals tinkering based on what type of experiment he was running. Greve modified his original design not only to make it generally more useful, but also to satisfy the parameters set by his experimental needs. The doppelküvette and the meteor planktonküvette maintain the horizontal water flow that Greve felt was more user friendly and had an added advantage for those interested in photography. The phytoplanktonküvette exchanged the sand bottom for the PVC to satisfy new demands for data collection but maintained other aspects of the original planktonkreisel. In essence, we see that even the original designer of a tank needed to modify that system almost immediately to fit the parameters of another experiment. In this way we can see that tank use was more about current need and local conditions than a basic form that worked for every person or situation.

Second, we can see that Greve was not necessarily working from precedent but instead, like most tinkerers, he developed his tanks based on his needs in the laboratory. Greve did not refer to Browne or Rees and Russell in this paper. In fact in the first paper for the planktonkreisel, he only cites one paper: a journal article describing the development of the sand filtration system for Helgoland tanks. In later papers he still does not cite any previous technological systems and only cites a single scholar regarding techniques for maintaining organisms. In this way, we see that we need not always trace a straight line from one tank to another but instead recognize that there are spaces where this knowledge might become combined but that technologies, even those that ostensibly do the same thing, are not necessarily related or built on previous iterations. Discovery in tanks is often disconnected from precedent.

COMBINING TECHNIQUES

In addition to his tank modifications, Greve was also incredibly interested in the process of understanding the diet and reproduction of these jellyfish species. In fact, the tank designs were to aid experiments on metabolism and reproduction. Unlike Browne, who largely worked on tank design without concern for developing methods of feeding and rearing, Greve did both. In his 1970 paper he gave detailed analysis of the food and life cycle of four species of common zooplankton. The longest lived was the *Pluerobrachia pilius,* a species of comb jelly known commonly as a sea gooseberry. This jelly survived 250 days in both the planktonkreisel and the doppelküvette. In the paper, Greve described both the food of the jelly and the feeding methods: "The food of young *Pluerobrachia pilius* consists mainly of unprotected larvae of benthic polychaets or molluscs. In the laboratory copepods, the main food source of adult individuals, exert a detrimental influence on young specimens."[31] He described the way that the species feeds, by passively floating into food. His understanding of their diet and behavior, coupled with the new tank designs, was successful in promoting sexual reproduction in captivity, resulting in a new generation of captive-bred comb jellies. Only 1 percent of his jellies produced by sexual reproduction survived into juvenile forms, but this was the first successful attempt at rearing jellyfish through their complete life cycle. Delap succeeded in raising and maintaining jellies from the moment of fertilization, but she never succeeded in getting her mature jellyfish to reproduce sexually. However, Greve was not the only researcher working on the caring labor needed to keep gelatinous zooplankton during this time.

Greve was part of a large community of international scholars using aquariums to determine the place of jellyfish in the marine ecosystem during this period. Between the mid-1950s and the late 1970s, scholars throughout the world built on Lebour's work to understand how and what jellyfish ate. The Russian biologist M. M. Kamshilov followed up Lebour's work with *Beroe* jellies. Kamshilov was the founder of the Laboratory of Plankton of the Murmansk Marine Biological Institute in Russia in 1952. His work at the marine laboratory was expansive, including surveys of the local fauna and studying the ecology of the Barents and Kara seas. Between 1955 and 1960 he performed a series of feeding experiments to test how jellies responded to prey. In addition to describing the process of eating and digesting in depth, Kamshilov also highlighted the important role of *Beroe* in the ecosystem of the Barents Sea. His work suggested that the carnivorous jellies changed the population dynamics of much of their prey species and were an important part of the ocean ecosystem.[32] Greve's work contributed to this discussion and was followed up by N. R. Swanberg's work on *Beroe* in the 1970s. Swanberg also placed jellyfish in aquariums and recorded their feeding habits over time.[33] In Miami L. D. Baker and M. R. Reeve used aquariums to monitor the influence of feeding on egg production in the hopes of understanding the cause of jellyfish blooms.[34] Other scholars used aquariums for observation of species but did not seek to keep them alive for longer periods.

A. K. Nagabhushanam, a biologist on the Isle of Mann, collected and watched northern comb jellies feeding in an aquarium. However, after four days they died, and he made no effort to maintain them by building a long-term tank environment.[35]

Most of the scholars working on the feeding of jellyfish in captivity did not seek long-term maintenance of the organisms. All the scholars mentioned above were catching their jellyfish each season and running feeding experiments to ascertain their role in the food web. None of the scholars, even Greve, matched the tank structure with feeding to extend the life of the jellyfish for the long term or to work to breed them. The work by Baker and Reeve did mention a tank with a small upwelling of air, and they were looking at reproduction, but this was not to try to get a generation of laboratory-born jellies but instead to monitor the rate of gamete production based on feeding. One researcher, Jed Hirota, did maintain two generations in captivity and matched both tank and feeding to a single species. Hirota, a researcher at the Scripps Institution of Oceanography in California, succeeded in rearing two generations of sea gooseberries in the laboratory using a combination of tank technology closely resembling Browne's plunger system combined with a feeding method developed by aquarists in Japan (although he did not cite either of these groups in his paper).[36] However, most of these scholars' tanks were built to work as close observation spaces in order to allow the observer full visual access to the jellyfish during feeding and to facilitate recording and photographing. Their experiments contributed to knowledge about the feeding habits and prey of jellyfish. But for a variety of reasons Japanese aquarists were interested in learning about long-term maintenance of jellies.

In Japan there were three groups working with free-swimming animals. The first group was fisheries biologists, and particularly those individuals interested in the aquaculture of lobsters. At the turn of the twentieth century, fisheries biologists in both the United States and Japan noticed a decline in lobster populations and began attempts to cultivate this economically important species in captivity. The earliest attempts at rearing lobsters failed because, while lobsters settle on the bottom of the ocean as they mature, as larvae and juveniles they float or swim in the pelagic water column. Any attempt to breed lobsters in captivity needed a technology that would simulate that environment. This group became particularly interested in Browne's plunger jar for this reason.[37] Others in Japan began working with jellyfish, including morphologists and physiologists at the Asamushi marine station and aquarists at the Ueno aquatic zoo. These researchers were some of the first to close the cycle of jellyfish in captivity (to rear jellies from sexually produced juveniles through the mature phase), and they also learned how to induce reproduction of species for quicker reproduction.

At the Asamushi marine station, Eturô Hirai and Yoshiko Kakinuma began investigating the life cycle of a variety of jellyfish in 1957. After collecting *Cladonema radiatum* (root-arm medusa) from the bay near Asamushi station, Hirai and Kakinuma placed the adult medusae in a glass container. In the morning, small planulae were recovered from the bottom of the glass and transferred to a smaller glass container, where they settled into polyps on the bottom of the dish. The polyps developed into extensive colonies and eventually budded into free-swimming ephyra. According to Hirai and Kakinuma, these jellies did not require continuous current (both Browne and Rees and Russell are cited in the paper), and the authors introduce a new form of feeding. Instead of replacing water to provide plankton, the researchers instead provided aquarium-reared *Artemia salina* (brine shrimp). Hirai and Kakinuma succeeded in rearing several species of jellyfish from planulae to medusae, including *Aurelia aurita* (moon jellies), by experimenting with brine shrimp, and they popularized this method of feeding.[38]

In her individual research, Kakinuma was particularly interested in investigating the effect of temperature on strobilation (the part of the life cycle when polyps bud into free-swimming ephyra) of medusae in captivity. In 1962 she experimented with both lighting and water temperature to induce asexual reproduction of captive jellyfish during the winter season (when these jellies would not naturally strobilate). Her work with Hirai combined with her own research on induced strobilization shed light not only on the natural life cycle of several species most common to Japan but also showed other members of the marine network a stable method for rearing jellyfish in the laboratory and producing mass numbers of jellies for either experimentation or, in the case of the public aquarium, exhibition.[39]

At the Ueno Zoo Aquarium in Ueno, Japan, Yoshihisa Abe and Michio Hisada became the first aquarists to develop methods to consistently close the cycle of moon jellyfish and to produce mature jellyfish for public exhibition. In 1965 Abe and Hisada began working to breed jellyfish in captivity. In 1968 they gave a presentation at the General Meeting of the Japanese Association of Zoos and Aquariums detailing their methods, and in 1969 they published a paper concerning those methods. According to the authors, the best method was a fairly simple one; Abe and Hisada placed early polyps in a flat aquarium with a sand filter bottom. During most months, polyps spontaneously strobilated, releasing ephyrae into the tank; when strobilation slowed or stopped, Abe and Hisada used

Kakinuma's method of light and temperature shock to induce it. The researchers, noticing the exhaustion of the ephyrae without current, sought to provide water movement via Browne's methods of mechanical movement. However, they found that the delicate zooplankton were too often torn apart or agitated by this movement and abandoned this type of current for a gentler form of convection via water movement and air bubbles.

Abe and Hisada fed their polyps and ephyrae with brine shrimp and their fully mature medusae with brine shrimp and chopped up clam and shrimp. All ephyrae and medusae were supplied with a gentle upswelling convection current made by air bubbles. One of the most important aspects of successful breeding of these jellies was their transfer during different periods of their development. The planulae and polyps were kept in a shallow aquarium with a sand filter bottom. As the polyps budded, the researchers transferred ephyrae to a tank with no sand bottom and a gentle current. That tank's water was not changed for thirty days, but brine shrimp were added through the air bubble system. As the medusae matured, they were then transferred to a bigger tank with a sand filter bottom and more current. The most important aspect of this paper is the emphasis on timing both technology and technique with the life cycle of the jelly.[40]

There were also a few physiology scholars continuing to try to rear jellyfish in the laboratory for medical use. In 1965 Dorothy Breslin Spangenberg, a neurophysiology researcher at the University of Arkansas Medical School in the United States also succeeded in maintaining and closing the cycle of moon jellies. Spangenberg's interest was in the development of the polyps directly after strobilation; she wanted to know the process of development of a polyp. She kept her mature jellies, derived from a single female collected in the Gulf of Mexico, in artificial seawater for over three years and fed them brine shrimp. She fed them twice weekly and moved them to fresh seawater after each feeding. Interestingly, Spangenberg cites Delap's research in her work, making her one of the only researchers to do so during this period.[41] Spangenberg's contributions are important in this field because they show continuity in the use of jellyfish in neurophysiology research and because her efforts demonstrate the isolated nature of tank work during this period. Her research was in the middle of the United States, and she was largely detached from the marine network. Instead, she was using a particular type of organism and tinkering largely on her own, away from the shore.[42] But unlike the isolated cases of husbandry development around the world, the core of the work on jellyfish rearing and maintaining during this period occurred in Japan.

After the earliest work with the plunger jar and rocking beakers combined with experimentation on rearing and feeding methods, the centers of knowledge production on gelatinous zooplankton fractured geographically and disciplinarily. This fracturing shows how much of tank work relies on localized understanding of laboratory needs and tinkering. In the sciences, we usually see a robust set of consistent citations in each paper as a field builds on itself. This is not the case with the development of jellyfish husbandry and tanks. Lebour is only cited by Kamshilov; Spangenberg is the only scholar who cites Delap. And in many cases, it looks as if aquarists are building on previous work without citing it. Hirota does not cite Abe and Hisada or Browne, even though his work seems to be closely aligned with theirs. This does not mean that aquarists are purposely underattributing craft knowledge, but instead it is an indication that tinkering does not lend itself to formal citations. Spangenberg did not cite any newer scholars on feeding or husbandry; reading Delap obviously gave her the confidence and information she needed to build craft by tinkering with her own system. This is seemingly the case with most of these scholars. They rarely cite widely but instead cite a limited number of scholars who performed similar craft in a slightly different manner.

These scholars also were not in direct communication with each other. While it is clear that groups in Japan were communicating, the rest of the community of tank users interested in jellyfish were spread far afield. In fact most of the citations are in marine laboratory journals that were not even widely available or translated into other languages. This lack of communication, both formally in citations and informally through travel, means that it took another decade for much of this tank craft to converge into the next solid advancements in jellyfish husbandry. It was not until the 1980s that both sets of knowledge—both the new tank structures that Greve described and the feeding and maintenance routines worked out around the world—combined at public aquariums in Germany and, eventually, the United States.

Technology and Technique Converge

Viewing jellyfish in captivity has become a common practice at public aquariums worldwide. The first major jellyfish exhibit containing a wide variety of jellyfish was displayed at the Monterey Bay Aquarium in 1992. This exhibit opened to great fanfare, and other aquariums soon tried to mimic the success of the aquarium using many of the technology and husbandry methods perfected, but not necessarily invented, at Monterey. The early 1990s saw the fractured research done throughout the globe on captive jellyfish husbandry combine into a coherent set of technologies and techniques for keeping a wide array of jellyfish species in captivity for extended periods. This combination occurred at the Monterey Aquarium, home of one of the best-known engineer/tinkerers in the aquarium industry, David Powell, and a relentless and systematic husbandry researcher, Freya Sommer. These individuals worked together to combine previous work in these areas and cemented some of the most basic craft of jellyfish husbandry used today.

However, the localized nature of jellyfish keeping remained after this crucial period. The combination of these forms of knowledge did not erase the need for a basic understanding of organisms, their specific diet, water chemistry, lighting needs, and the technological and husbandry tinkering required to provide those needed parameters. This section shows the power of public aquariums to combine information from a wide network of users into operational forms as well as to move that knowledge into other areas of the network, including the hobbyist community. In addition, it shows how changes in technological systems, the actual tank designs, become incremental and increasingly more site and species specific. Many of the changes in tank design at public aquariums during this period were created to make tanks functional as exhibitions both capable of sustaining life and aesthetically pleasing to look at from certain angles. At the Monterey Bay Aquarium, tinkering with kreisel tanks started from basic kreisel specifications, but subtle changes were made in size, shape, and water intake to keep a wider range of species. In contrast, the husbandry techniques and knowledge of environmental requirements and specialized diets needed for each species of jellyfish remains a growing field. In this way, it becomes clear that the work of keeping jellyfish in captivity today requires both a stable technological system, an understanding of the known essential elements for individual species, and continuous tinkering to interpret that knowledge for a local system.

As the second generation of craft knowledge for jellyfish keeping became more widespread, public aquariums took interest in Ueno's displays and sought to emulate them. For instance, in 1986 the Dusseldorf Aquarium in Germany became the first public aquarium to combine Ueno's techniques for rearing and displaying moon jellyfish with the kreisel tank. It was reported that "For optimal exhibition, however, the Dusseldorf method requires a rather large planktonkreisel, which cannot be constructed in traditional aquarium tanks. If the diameter of the kreisel is too small, the water circulation and the narrow area make it almost impossible to observe all the typical jellyfish movements."[43]

Three years later, the Berlin Aquarium worked out a system that allowed them to keep moon jellies in a traditional square aquarium by adjusting the water movement to keep jellies away from the filters or sides.

Both of these aquariums succeeded in building on Ueno's success with moon jellies and set the stage for other aquariums and researchers to take interest in this growing trend.

In 1990 William Hamner, a plankton researcher at the University of California, Los Angeles, in the United States, introduced his updated design for the kreisel tank. Hamner based his design on Greve's meteorplanktonküvette but chose to retain the name of kreisel "because of the circular water rotation in the aquarium and also because the term is both expressive and attractive."[44] This tank is not significantly different in function from Greve's design; water circulates in a vertical vortex instead of horizontal. Differences included inlet and output for water at the top of the tank (instead of the bottom), input jets that pushed outgoing water across a net, and tank lighting by dark field illumination. The first two alterations to Greve's design stopped organisms from getting caught in the filtration system and stopped air from entering the system anywhere but the water current. The final difference of lighting shows how Hamner shaped the tank for specific observation requirements. The entire system was darkened except a viewing section in the front of the aquarium, meaning that it was easier to make out the shadows of the organisms inside. Because Hamner was interested in small gelatinous zooplankton, it makes sense that he made this particular change.

Another important aspect of Hamner's design is the use of acrylic for the entire system. Because the central tank had to be smooth and circular in shape, it was easier in construction to use acrylic for this piece. However, there were no known bonds at this time to stick acrylic to glass, meaning that Hamner constructed the whole system out of plastic, something that had not been done before. Hamner's design, built in the mid-1980s, is important for these changes to Greve's system but also because in 1988 he loaned his tank to David Powell, the chief engineer at the Monterey Bay Aquarium. At the time of his publication in 1990, this kreisel tank design had been used by Powell at the aquarium for two years.

JELLYFISH DISPLAYS AT THE MONTEREY BAY AQUARIUM

Following the success of displaying *Aurelia aurita* at the Ueno Aquarium, the Monterey Bay Aquarium joined the Dusseldorf and Berlin aquariums in seeking to set up a yearlong display of these jellies. Abe and Hisada shared not only their husbandry knowledge with Monterey Bay's chief jellyfish keeper, Freya Sommer, but they also shipped polyps from their moon jellies to Monterey to start the exhibits. Sommer began her tinkering toward displaying jellyfish at Monterey in the mid-1980s. The first year-round exhibit of moon jellies began in 1985. Sommer described her process of maintaining a variety of moon jellies, including the Japanese specimens and those collected locally in California waters. According to Sommer, the California-collected jellies developed normally, being fed daily with brine shrimp. However, the Japanese medusae were stunted in their growth and failed to thrive with this diet. Sommer tinkered with the diet by feeding the jellies brine shrimp soaked in Selco, a special combination of omega-3 fatty and amino acids. The combination of brine shrimp and Selco had previously been developed for some of their captive fish stocks (see chap. 5), but the jellies thrived on the diet as well. Unlike previous authors who suggested a combination of brine shrimp and chopped clams and shrimp for adult jellies, Sommer felt that the brine shrimp combination was sufficient and that other food was both inconsistently available and difficult to clear from the tank after feeding.[45]

The tank designs outlined by Sommer do not differ greatly from those previously described by Hamner. Hamner was a consultant on the work being done at the Monterey Bay Aquarium and his work was equally influenced by David Powell, the director of husbandry at the aquarium, during this period. Powell is a well-known aquarium exhibit designer, having created some of the most recognizable public aquarium exhibits—including the iconic kelp forest and wharf piling.[46] Sommer and Powell worked together at the aquarium to display every aspect of the jellyfish life cycle. Initially, they separated the groups into polyps and ephyrae in one tank and full-grown medusae in another. However, the polyps ate newly budded ephyrae, resulting in a diminished population and leading to separate tanks for these life cycles. The polyps were suspended on an acrylic plate in a square tank with minimal water movement. The aquarium displayed ephyrae in a small kreisel tank with minimal lighting and a completely dark background and larger jellyfish in a scaled-up kreisel with the same scheme.[47]

Sommer's attempts at maintaining moon jellies collected from several locations revealed an interesting problem for individuals hoping to follow her guidelines for displaying these organisms: animals collected from different localities responded differently to the same exact aquarium conditions. Moon jellies have a wide global distribution; they occur in temperate, arctic, and tropical seas. Working with this species collected in Japan and California, Sommer found that the different groups spawned at different times, had different life expectancies, and had varied body morphology. In particular, the Japanese specimens remained smaller and lived almost a year less than their California counterparts. In addition, while moon jellies occur in tropical waters, neither group could exist outside the parameters of the water temperature in which they were initially caught. Sommer used this comparative information to suggest that while there are guidelines that can be followed for maintaining jellies in captivity, they must be plastic enough to take into account the differences not just in species but in the location where specimens were caught and kept.

After the success of the moon jelly display, Sommer and Powell sought a larger exhibit of a wider array of species. In 1992 the special exhibition "Planet of the Jellies" opened to great fanfare. The exhibit pushed jellyfish husbandry forward. It contained seventeen species of jelly, including Japanese moon jellies from Abe and Hisada and sea gooseberries from Hirota in Washington State. The tinkering required to develop craft for keeping new species resulted in new understandings of many species. Powell recounts a moment when Sommer succeeded in closing the cycle of another species. "Not only was this an outstanding husbandry success, but it was also a scientific breakthrough. This discovery of the previously unknown life cycle of the West Coast *Pelagia* may eventually result in this jelly's being reclassified and placed in a different genus."[48] Sommer's findings contributed to a 2002 paper that sought to clarify the phylogeny of *Pelagiidae*; Sommer's work showed that there were marked differences in coloration of certain members of the family in polyp stages.[49]

What is so impressive about this display was the variety of tank designs, water, and feeding of each separate species. For the initial 1985 moon jelly display, there were three separate tanks and a specialized diet. Multiply this tinkering by seventeen and you will get the full idea of the amazing nature of this exhibit and the work of Sommer and Powell. Unlike moon jellies, the diet of other species is more specific. As Lebour and Delap found at the turn of the twentieth century, many larger jellies prefer eating other jellyfish or fish. In her publications on this exhibit, Sommer detailed all the new foods found to be useful to keeping jellies. In addition to brine shrimp enriched with Selco, the sea gooseberries required gelatin and wild plankton. Often, food depended on maturity and life stage. For instance, small, newly formed polyps of some species could not eat the comparatively large brine shrimp and were fed rotifers; in some cases, even rotifers proved too large, and oyster larvae were substituted. As the polyps developed, they could then be switched to the brine shrimp. Larger medusae, including the ever-popular sea nettle, were fed immature moon jellies and frozen krill. These diets had to be timed to align with the life cycle of the species and modified if the individual species showed signs of distress.[50] In addition to the development of feeding schedules for the exhibit, there was a wide array of tanks, including a new modification to the traditional Greve/Hamner kreisel.

For the most part, the modifications of the Hamner kreisels were done to make feeding and scaling up easier. The main changes in the kreisels were aesthetic. Tanks were lit from the sides, and the back was a translucent blue backlit with a fluorescent light. This lighting caused an "endless blue" effect that made the tank appear larger than it actually was. The biggest finding was that the kreisel design could be modified to keep a truly wide array of species, many that did not need the same rate or directionality of current. *Cassiopea xamachana* (upside-down jelly) do not need a kreisel tank because they are a mainly stationary species that live in relatively stagnant water; their tanks were sand filtered tetrahedrons with rectangular bottoms and heated water. Other species, including *Polyorchis penicillatus* (bell jelly) are small and could be kept in pseudo kreisels, which work on the same theory of traditional kreisels but are modified from traditional rectangular tanks.

The main change to the kreisel tank system for this exhibit was what became known as a "stretch kreisel" or "Langmuir kreisel." In physical oceanography, a Langmuir circulation is a set of countercirculating currents on the water's surface aligned with the wind. Imagine a set of countercirculating drains on the surface of the ocean (a gentle circling). The stretch kreisel replicates this dual current with inlets for water at opposite ends of a long kreisel tank. As the water meets in the middle of the tank, water is pushed down by the competing currents and back toward the opposite side of the tank, where the water is then pushed upward as it hits the sides of the curved tank. Sommer explains that this type of tank is good for jellies that swim into currents and require slightly more movement, such as the *Pelagia colorata* (also known as the *Chrysaora colorata*, or purple-striped jelly). All tanks had separate water from the main aquarium tanks, and each tank had separate intake systems to account for changes in chemistry and temperature requirements.[51]

Most of the publications resulting from the 1992 exhibit revolved around the exhibition and gave less information about the breeding of these species. Sommer's earlier publication regarding moon jellies talked about spawning and life cycles, but the focus on exhibition left little room to discuss the difficulties of continuous breeding of all of these jellies. It is clear that many of these species were short lived in their tanks and had to be collected several times. However, many jellies survived for years in these tanks. For this ability to combine the technology and husbandry into a workable system, Monterey Bay Aquarium and Sommer's publications on her jellyfish research became the touchpoint for both the public aquarium and the hobbyist community's growing interest in displaying and breeding jellies.

Jellyfish Keeping Spreads throughout the Marine Network

Powell and Sommer's success with jellyfish husbandry was a result not just of their own tinkering but of the combination of far-flung craft knowledge developed in the previous generation of jellyfish keepers. As stated previously, Jed Hirota and Abe and Hisada shared information and specimens with Monterey. Hamner shared his tank designs with Powell. But there were many others who shared expertise. Soon after the exhibit opened, Kazuko Shimura, the jellyfish expert from Enoshima Aquarium, visited Monterey. Although Shimura did not publish much of her work, she was known as the foremost expert in jellyfish husbandry in the public aquarium community, having maintained the most species in captivity to that date.[52] During her visit, she noticed that the *Palau* jellies needed more light to photosynthesize and expressed this to Powell and Sommer. This form of knowledge exchange was integral to the success of the exhibit.

Monterey also exchanged craft knowledge with other research scientists. As stated in the previous sections, academic researchers from a wide range of disciplines used tanks to observe and experiment with jellyfish. But while this research resulted in knowledge about essential elements of jellyfish environments, the researchers did not seek to maintain these jellyfish for the long term. However, this does not mean they were not interested in the outcome of these husbandry experiments. Powell recalls,

> Although the lives of planktonic creatures had been widely studied by a number of people in the academic world, few researchers had succeeded in keeping these animals alive for any length of time. Some of these scientists—most notably Claudia Mills of the University of Washington, Sid Tamm of Woods Hole Oceanographic Laboratory, and Ron Larson of Harbor Branch Oceanographic Institute in Florida—were excited about the work we were tackling and enthusiastically shared their knowledge with Freya. Their encouragement and support were a great help.[53]

The Monterey Bay Aquarium became the epicenter for craft exchange from a wide network of aquarium users who both shared and received knowledge resulting from new tinkering.

Today, scholars continue to strengthen the links between jelly tank craft and knowledge production about the sea. In 2017 Rebecca Helm and Casey Dunn published a paper detailing the use of certain chemical compounds to induce strobilation in a wide range of jellyfish. According to Helm, she began this work when she became frustrated with her inability to find a species that would allow the study of strobilation. Instead of continuing to search for a jellyfish that strobilated readily in the laboratory, Helm turned to the literature and found the history of compounds used by other aquarists, including Spangenberg, to induce strobilation. After much tinkering, Helm found that indoles induced strobilation in a wide array of species. She wrote in a blog post that "now, rather than traveling halfway around the globe, scientists can add a couple drops of a special compound to their polyp tank, and have jellyfish to study in under a week! I hope this work will be helpful to many different kinds of jelly-lovers, form biologists to aquarists and beyond."[54] The research Helm and Dunn published not only made it possible for experimentalists to reproduce jellyfish in captivity but it also explained triggers for strobilation in the open ocean.[55]

But while jellyfish keeping spread in the public aquarium and research communities, the entrance of the hobbyist community into jellyfish keeping came much later in this narrative than in other forms of tank development. In the case of photographic tanks (chap. 2) and reef tanks (chap. 4), hobbyists were pioneers in tank and method development and dissemination. In jellyfish husbandry hobbyists did not serve as pioneers in the development of the kreisel tank. There are multiple reasons that this might be the case, many of them logistic. Even though Spangenberg kept her jellies in Arkansas, collecting and transporting jellyfish from water sources to inland locations remains a particularly delicate process. Every other display of jellyfish was done near the water from which the jellies were collected. Sommer highlights the local collecting she performed for the Monterey Bay Aquarium exhibitions and describes it thus:

> Local collecting is done with a seventeen-foot Boston Whaler, usually within a mile from shore in and around Monterey Bay. The speed of the Whaler permits coverage of a great deal of area in the search for "slicks," glassy patches on the ocean surface caused by a combination of wind and current, where jellies are most likely to be seen. Jellies may be collected by dipping from the surface, or by snorkeling or SCUBA diving. Some of the hardier species may be collected in fine-mesh or knotless-mesh hand nets. Ideally, the animals should remain supported in water as they are brought onto the boat.[56]

The difficulty with shipping and the diet and space requirements needed to breed jellies in captivity meant that aquarists could not rely on the ornamental hobby retail companies to provide them with live specimens; the only way to get your own jellies was to collect them yourself, usually from a boat that could go far enough, fast enough, and long enough to find jellies. Unlike ornamental fish, jellies do not congregate on reefs or other structures but instead float in large groups on currents. The difficulties involved in procuring specimens, on top of the difficult process of keeping them alive in captivity, meant that hobbyists only entered this narrative in the early 2000s.

Monterey Bay Aquarium embraced their role as the epicenter of jellyfish husbandry, spreading craft knowledge to academic aquarists, public aquarists, and hobbyists. For public aquarists, the information shared at professional meetings and through publications helped build massive exhibitions throughout the world. In the 2000s the Monterey Bay Aquarium began hosting a Jelly School for aquarists to learn tank craft. It is described as "a three-day workshop where colleagues from around the world gather to share in cnidarian knowledge-building."[57] In addition to this work with the aquarist community, jellyfish husbandry experts have helped jump-start the home aquarium hobby. Chad Widmer, formally the senior aquarist at the Monterey Bay Aquarium working with jellyfish, wrote a book titled *How to Keep Jellyfish in Aquariums* (2008). It is one of very few books on jellyfish husbandry, possibly because the majority of information is exchanged in online forums and through hobbyist meetings.

The conversations in the hobbyist community show a continued struggle with local conditions and a need for a "blue thumb" when tinkering with both the system and specific needs of jellies. In his book, Widmer includes a section titled "Assumptions about the Reader." In it, he states that he assumes any reader of the book is not a beginner and that "keeping jellyfish is for advanced aquarists, not because the work is difficult or hard to understand but because the life-support system requires daily attention and appropriate responses from the aquarist for success."[58] Additionally, he stresses the difficulty of mastering the craft: "I also assume you realize that there is a learning curve and that you are unlikely to be successful with your very first batch of jellies but will soon become good at it."[59] Widmer's book offers loose guidelines for keeping a wide variety of species, but he is clear that these are just guidelines. In many sections he offers at least three possible reasons for something to go wrong and as many ways to fix it. The advanced aquarist is supposed to combine previous experience with knowledge of their system and specimen and apply these general guidelines judiciously. The expectation that jellyfish keeping is laborious and will result in initial, and possibly consistent, failure is something that Widmer acknowledges and, indeed, expects that his audience already knows.

The craft nature of jellyfish husbandry becomes clearer when seen through the eyes of first-time hobbyists forced to reckon with the inability to find clear, quick fixes for their problems. This was the case in the opening of this chapter; the initial investors in Andon's jellyfish tank were unaware of the knowledge base required to succeed at keeping these organisms in captivity. While many were willing to learn by joining online communities and reading Widmer's book, others quickly lost interest in the aesthetic beauty of jellyfish because of the difficulties inherent in keeping them looking so beautiful. The Andon tank was the first kreisel tank to be patented in the United States. However, the patent did not mean that no technical knowledge was required to run the tank. Many users found that the intake valves needed to be moved or that the engine was too strong or weak to provide the optimal current for their jellies. Others have learned to moderate their seawater mixture (which comes with the tank) and lighting schemes. In addition to these tank fixes, internet boards are clogged with information on feeding schedules and questions about nursing injured jellies. Many keepers, initially seeing the beauty of a streamlined tank for their living room, have multiple tanks to nurse injured adults and to keep ephyrae and polyps. In essence, you cannot be an amateur hobbyist in the jellyfish hobby.

Conclusion

The history of the kreisel tank is one that shows the movement of technology and technique through the marine network in a nonlinear path. Instead of moving from one group or industry to another, different forms of knowledge were developed in geographically and disciplinarily diverse locations simultaneously. In the case of the kreisel tank, the important node of the network is the public aquarium space, where continuous exhibition forced researchers to combine various knowledges from throughout the network, which resulted in more standardized technological systems paired with consistent techniques. Through the combination of craft developed throughout the marine network, public aquarists developed the first set of recommendations for keeping a wide range of jellyfish in captivity. These recommendations, capable of being used as the basis for more tinkering based on localized conditions, were then passed back into the marine network.

The history of the kreisel tank also shows the difficulty of tracing both forms of craft knowledge production through the network. Throughout the development of jellyfish husbandry, the citations for technological development are more consistent and traceable. When you follow each paper, you can easily track back to the previous tank iteration through citations. This is not the case with techniques. Researchers responsible for knowledge about husbandry techniques are rarely acknowledged in publications, and the line of acknowledgment for many findings is often broken or lost. This does not mean we cannot track the spread of techniques, but they are often invisible if you are trying to trace them in publications. The use of brine shrimp as feed clearly spreads from Ueno to Monterey Bay and then throughout the network. The addition of Selco to the brine shrimp is a common practice today mentioned in nearly every publication or board touting advice. But Abe and Hisada are not credited for their use of brine shrimp, nor is Sommer for the addition of Selco. In essence, once techniques enter the network, they are treated as common knowledge. In this way they are undervalued as not "discoveries" or "scientific." Even though techniques tell researchers in the network quite a bit about the physiology, life cycle, and behavior of the organisms that they keep in captivity—and help maintain the organisms in a condition that allows experimentation and observation—the knowledge is deemed too common to cite.

The development of jellyfish keeping in the hobbyist community, and especially the sharing of "tips and tricks" for keeping jellyfish, has further shifted the recognition of technique attribution. Of course, there are still attributions and citations in formal papers, but if any of that information was learned or passed through the hobbyist or aquarist community, the citation makes it seem as if it is somehow new knowledge instead of something that has circulated throughout the community for some time. The addition of a new community of knowledge producers makes it even more difficult to trace and attribute the development of techniques to a single person or group of people. Moreover, it is difficult to distinguish where knowledge develops when group boundaries are blurred. Similar to the overlap between photographic and fish hobbyists in chapter 2, the line between "hobbyist" and "professional" is one that is usually drawn around space or job description, not person. An individual can be both a hobbyist and a professional aquarist or an academic researcher and hobbyist. The blurry line between work done at home or in the laboratory and the murkiness between what you've learned from the community online and then applied in the laboratory makes tracing technique production through this community challenging.

The interplay between field observation and tank craft resulted in both enhanced knowledge of wild jelly populations and new systems for helping captive jellyfish thrive. Before jellyfish were able to be maintained in captivity, little was known about their life cycle, hunting and eating habits, or the environmental variables that stimulate growth and sexual development. The act of maintaining these organisms in captivity provided information about their basic biology and natural history. Finding the "essential elements" of the environment through tinkering helped researchers build knowledge about wild jellyfish. But finding these elements also demanded technologies and techniques for simulating multiple forms of complex water circulations and feeding grounds. As aquarists developed understandings of the importance of water movement and feeding to maintaining specific organisms, oceans under glass became more varied and expansive, with thousands of tanks throughout the world simulating niches for specific species and life cycles. As we will see in the next chapter, increasing the number of species kept in a single tank built directly on the craft of jellyfish aquarists. But unlike kreisel tanks, building increasingly complex ecosystems requires the aquarist to make even more choices about what they believe should and should not belong in a tank. The more complex a tank becomes, the more it becomes a simulation of the natural system, and the less likely it is to model that system. The resulting tank is an imaginary ocean under glass.

4

Reef Tanks

Building Ecosystems under Glass

In August 2017 I found myself sitting quietly in front of a large tank at the St. Lucie County Aquarium in Fort Pierce, Florida. The tank, a simulation of a Caribbean coral reef that can be viewed from three sides, takes up almost an entire room of the small aquarium. There are several species of coral that make up a huge mound in the middle of the tank. At the top, endangered staghorn coral crown this structure as it reaches up to the lights suspended above the tank. The water is slightly cloudy, looking almost milky, as it shifts back and forth in a simulation of wave movement caused by a bucket system that dumps water into the sides of the tank every few minutes. Each day I chose a different area on which to concentrate. One day I focused on the colorful free-swimming fish darting throughout the tank. The next I tracked the crustaceans crawling into the crevices of coral. But this morning I was focused on a small, yellow fish burrowed into the sand substrate in the corner of the tank. This blenny was digging furiously, darting into his hole every time another tankmate approached. In between interviews and writing my notes, I visited this industrious fish to see how he fared. The head aquarist told me that the blenny had moved his burrow several times because he was being picked on by others. He had, hopefully, finally found a good place to settle. Throughout the day, I followed the drama of this tiny fish, and later, as I prepared my fieldwork notes, I thought how *real* and *natural* the behavior of this blenny seemed to me. Watching him move sand all day and knowing that this was part of a longer life course of movement within this environment brought to the fore the realism contained in these tank simulations.[1]

I went to Florida to study the aquarium's reef tank because it has a long and somewhat controversial history in both marine science and aquarium hobbyist communities. The main reef tank in St. Lucie was built by Walter Adey for the Smithsonian Institution's National Museum of Natural History (NMNH) in 1980. Adey began building experimental

FIGURE 4.1 Adey tank at the St. Lucie County Aquarium is a model of a Caribbean tank. The filtration system developed by Walter Adey is supposed to maintain needed nutrition for the reef and its inhabitants but results in murkier water quality. Courtesy of the St. Lucie County Aquarium.

tanks in the basement of the NMNH in 1978. He conceived of these systems as *microcosms*, or enclosed ecosystems that functioned so similarly to natural ecosystems that experiments performed in them would reveal information about their wild counterparts. Adey and Karen Loveland built a small microcosm of a Caribbean coral reef in the basement, and two years later they built a larger tank in the NMNH's marine display hall. The NMNH used the tank to educate the public about reefs and made it available to scientists for experimental purposes. The tank was not an unqualified success at achieving either of these goals, and in 2005 it was transferred to St. Lucie County Aquarium. Today it is no longer used for formal experimentation and functions solely as an exhibit. It is still an impressive display of tank craft, but it is made even more interesting when the history of the piece is revealed.[2]

Reef tanks are replications of marine reef systems. The reef tank is supposed to be a complex environment that contains a variety of organisms that not only live in that environment but contribute to the ecological workings of that space. Adey and Loveland describe reef tanks as "the most complex ecosystems in the sea, coral reefs, in microcosms of a few cubic meters, behaving chemically as wild reefs, and having a biotic diversity per square meter exceeding that known for the wild."[3] A reef tank is a combination of invertebrate and vertebrate organisms, plants, and bacteria meant to mimic the working of a wild reef. The intricacy of the reef tank is mirrored in the difficulties developing craft to maintain these systems. The tanks are filled with a variety of organisms requiring a careful balance of water chemistry, light exposure, and food. Developing tank craft for this intricate system is no small feat.

The aquarium network built the first reef models directly next to the source of their specimens, easing the transition from the open ocean to oceans under glass. Hobbyists and public aquarists built the first reef tanks close to the water throughout the Pacific. Public aquarists in New Zealand, Hawaii, and Indonesia all built tanks that were mere steps from the locations where they collected their coral. While there are many problems involved with maintaining a coral tank, the first issue that aquarists had to overcome was the movement of delicate corals from the bottom of the ocean into a tank. These early reef aquarists determined that success with coral movement required quick work to reduce stress. Adjacent locations allowed aquarists to source delicate coral nearby and put it directly into a similar environment. To make the tank as similar to the native reef as possible, the aquarists used unfiltered seawater and natural sunlight for their tanks. While many tanks were initially built relatively close to water sources, the first reef tanks were not separate from these water sources but connected to them via water pumps bringing local water into the tank. In addition to using unfiltered water and natural sunlight instead of artificial sources, building tanks so close to their collection allowed aquarists to use the local reef to compare and contrast with their tanks, making it easier to gauge the success or failure of their corals and to try to understand why either occurred. The wild reef became a reference for the tank reef, serving as a handy resource for working out the essential elements, including predator and prey relationships and seasonal changes of reefs.

While the earliest tanks were built in the Indo-Pacific because of the difficulty in moving coral, advancements in shipping corals and the movement of a wide variety of corals and fishes throughout the marine network jump-started worldwide reef keeping. By the mid-1980s, "mini reef" keeping was popular in Germany, with pockets of aquarists and hobbyists keeping reef tanks worldwide. The movement of corals was matched by the movement of knowledge. By the late 1980s there were several hobbyist magazines, including *Tropical Fish Hobbyist* and

Freshwater and Marine Aquarium (*FAMA*), sharing tips and tricks on reef keeping. By the early 1990s *FAMA* had a special section of the magazine dedicated to reef keeping, with a monthly question and answer column by Julian Sprung.[4] In addition to the magazines, reef-keeping manuals began to appear on the market in the late 1970s with increasingly long and beautifully illustrated texts appearing consistently into the 1990s. The internet opened new avenues for communication and circulation of materials. Message boards and online magazines such as reefkeeping.com and Reefs.com have allowed even more communication among communities starting in the early 2000s.

Because most aquarists worked with similar coral and fish and communicated through the same channels, there was open exchange of tank craft throughout the reef-keeping network. Unlike the photographic and kreisel tanks already discussed, tracing craft development and debates in the reef tank community is somewhat easier. Almost all magazine articles, online boards, and hobbyist guidebooks contain similar historical timelines regarding the development of reef tank craft. Whether you are reading an online board, listening to a conference talk, or reading an academic publication, Lee Chin Eng is cited as the first individual to consistently maintain a coral tank. Lee, a Chinese public aquarist working in Indonesia, published a paper on his method in 1961. This agreement on historical timeline means that reef tank keepers (referred to as "reefers") share acknowledgment of craft and the people they consider important to its development. One board on nano-reef.com in April 2020 suggests that knowing the history of the craft can inform current practice. One user going by the handle mcarroll states, "Seems like the methodology has been lost on the last few generations of reefers. Even back in 2017 nobody was talking about the Berlin method anymore. *Almost* nobody uses live rock . . . *almost* nobody even recognizes that live rock is better than what came before it."[5] Throughout the development of reef craft, the community has maintained close communication through agreed-on history about the development of their craft.

This open communication resulted in several clear debates about the nature of reef keeping. One major debate in the reef aquarium community is the question of how to achieve the most "natural" reef system. Nineteenth-century aquarists often touted the "balanced aquarium" as a system that demonstrated the perfection of nature. The chemist Robert Warrington's interest in aquariums involved using flora and fauna in the tank to achieve a balance that meant that filters were not required to sustain life.[6] The balanced aquarium is extremely difficult to achieve, but the debate about building balance into a system is widespread in the reef tank community. The earliest reef tanks built near the water contained recently collected coral and fish and used unwashed rocks (known as "live rock"), sunlight, and unfiltered seawater directly from the ocean to mimic the balance of the natural reef. Lee Chin Eng called this "Nature's system" and felt his success with a reef tank was achieved by following God's plan for nature.[7]

As reef tanks moved farther from the seaside, designs for new systems still sought to replicate the natural reef environment found in the open ocean, but debates began about how to achieve a balanced system. Aquarists hoping to replicate reef ecosystems debated the importance of water chemistry, especially the importance of added elements, and the role of algae and other floral components. Building on Lee's process, three different systems in Berlin, Monaco, and Washington, DC, arose that sought to isolate the essential elements of the reef ecosystem with varying focus on trace elements, water movement, and methods for filtration. While most aquarists were working with similar tank inhabitants, the debate about the importance of any given essential element to the larger function of the reef resulted in drastically different tank designs. The term *natural* is problematic here, as no tank simulation of a coral reef could replicate so many variables in such a small space. Reefers most often use the term *natural* to refer to a tank that needs minimal aquarist interventions to operate. Examining how they used the term tells us much about the way that aquarists think about their goals and role as tinkerer.

These debates featured questions about how much these tanks could mirror actual reef ecosystems and what they could tell us about wild reef systems. The rise of ecological and environmental sciences in the twentieth century coincides with the development of reef tanks throughout the world. The growth of laboratory simulations of environments, called *microcosms*, influenced the reef tank community. Jean Jaubert and Walter Adey, both academic researchers and tank hobbyists, tinkered with reef tanks, trying to develop systems that would serve as experimental technology and inform their marine research. Adey not only built tanks for the NMNH but he also worked with groups in Queensland, Australia, and the Biosphere 2 group to build large microcosms that he hoped would serve as experimental systems for a wide range of questions about reefs.[8] Other reefers, including commercial hobbyist Richard Ross and public aquarist Bruce Carlson, have called for more research scientists to use reef tanks for their studies and pointed to the importance of these systems for understanding reef ecosystems. But while the reef tank community has been vocal in the belief that these systems can be useful to developing a deeper understanding of reef ecosystems, the uptake of these tanks into formal academic knowledge production has been slow and unsteady.

Unlike the previous tanks discussed, there are purposely built-in unknowable elements of the reef tank. The most successful reef tanks seek to replicate the natural environment by including elements such as live rock and algae. "Live rock" refers to a rock that is placed in an aquarium without first treating it to remove any traces of algae or organisms from it. Using live rock in aquariums introduces a wide variety of unknown elements to the tank, and while early reef aquarists insisted this was required for success, the use of live rock also introduced new difficulties not heretofore seen in captivity. Live rock contains "good hitchhikers" for the tank, including algae and organisms, but there are also commonly "bad hitchhikers" on rock, organisms that are native to reef environments but destructive to captive systems. The Atlantic Reef Conservation website has a page dedicated to the identification of many hitchhikers, saying that "99% of these hitchhikers and reef critters are a valuable addition to your aquarium. This species identification guide will cover all these beneficial critters as well as the 1% that make up the bad hitchhikers and reef tank pests, as well as how to easily remove one should you be part of that 1% who get a hitchhiker pest that is not beneficial."[9] If you use live rock, you will introduce new, and often completely unforeseen, elements into your tank, creating a system that evolves somewhat organically in each individual tank.

Because of these unknowable elements and the need to control variables in experimental biology, coral microcosms used by biologists differ from those built by earlier reefers in the network. Experimental microcosms do seek to replicate some of the essential elements of the ecosystem, but they contain less diversity and unknown elements than most public aquarium and hobbyist reef tanks. In a 2017 paper about the challenges and opportunities found when working with coral microcosms, Patrick Schubert and Thomas Wilke suggest that tank parameters—including water temperature, chemistry, and hitchhikers—be as controlled as possible for experimentation. But the authors state that these tanks can add much to field observations regarding coral. "Microcosms enable the manipulation of a single or few variables, and to compare the effects on organisms over time against control conditions. However, unlike natural systems, microcosm experiments are an abstraction from reality, and no single setup might explain the complex impact of global change on populations, species, and communities. Instead, each setup may help answer a specific question. Besides generating such specific knowledge, microcosm studies can also help develop theories and meaningful policy implications."[10] Today, microcosms are used to isolate and control variables to test their effect on reef health. Across the causeway from the St. Lucie County Aquarium, a lab group led by Valerie Paul at the Smithsonian Marine Research Station uses open-air coral tanks to study the influence of a variety of variables on coral disease growth and spread.[11]

While microcosms may simplify the reef tanks developed by the larger reefer network, they owe their craft to that community. According to Schubert and Wilkie, the use of microcosms is relatively new, dating back only thirty years. Their citation for this history is a NOAA technical memorandum written by T. C. Bartlett in 2013. Bartlett's citations regarding tank structure, building, maintenance, and internal parameters are drawn from the hobbyist and public aquarist communities. He cites websites such as AdvancedAquarist.com and Reefkeeping.com, as well as popular reef-keeping manuals by Charles Delbeek, the curator of aquatics at the Steinhart Aquarium in California, and Julian Sprung, a commercial hobbyist.[12] While Bartlett's report does not indicate the broader history of reef tank development and use, the current craft used by academic biologists is a direct outgrowth of that history.

This chapter follows reef tank craft from the 1950s to the present to understand how it has developed and spread. While these tanks have become increasingly important to academic science in the last two decades, the craft has developed largely outside of the academy. Unlike the networks involved in the previous two tanks discussed, the network involved in the initial development of reef tanks was primarily made up of hobbyists, commercial aquarists, and public aquarists. The few academic researchers participating in the community did so as hobbyists, tinkering with tanks to satisfy their own goals and not necessarily those of their research work. The close communication between hobbyists, commercial groups, and public aquarists resulted in a robust conversation about not just the craft of the tank but also the philosophy of tank keeping. This community used conversations about craft to debate questions of how much of the natural environment could be simulated and how much it could be controlled. These conversations offer insight into how important these nonacademic communities are to both the development of tank craft and the philosophy of simulation creation.

Also important in this history is the way that craft travels through the marine network. For other tanks, we have seen knowledge couriers that straddle different communities, such as William Innes and the photographic tank and the importance of institutions as spaces for craft transfer in the case of the kreisel tank. In the case of reef tanks, we see a new form of the movement of knowledge, from the hobbyist and public aquarist communities into the academic community via academic researchers who are also hobbyists. The advancement of craft in reef tanks comes primarily from the public aquarist and hobbyist community. Individuals who straddled the academic research and hobbyist communities carried the information produced between those groups into the academic community. Academic researchers such as Walter Adey developed craft as hobbyists and participated in that community. Adey exchanged craft with hobbyists and aquarists, and his tinkering was shaped by the conversations occurring within those groups. He produced knowledge that fit the requirements for participation in hobbyist communities and developed a tank that was both scientifically useful but also became a long-lived public aquarium exhibit. But he also maintained his presence in the academic research community, and his work with reef tanks resulted in the wider circulation of the craft. This form of craft circulation is one we have not examined yet in this book, and it is important for thinking about how integral hobbyists and public aquarists are to conservation goals in the climate change era.

Today, coral reefs are in peril, and reef tanks are an important tool in understanding and stopping their decline. Tracing the history of reef tank craft shows how these tools developed and the importance of communication between the groups that built them. This history shows active debates about the nature of tank craft and the perceived relationship of tank-bound reefs to their wild counterparts. Those tinkering with these tanks did so while actively debating what they wanted the tanks to *be* and what they wanted to *do* with them. The conversations surrounding reef tank craft shows us how important various groups of tinkerers are, not just in technological design but in the conception of and the debate about the very nature of these tools. All the tinkering and debate has led to simulations so realistic that I could have sat watching that blenny build his nest for days.

Early Interest in Coral

FIELD RESEARCH

There are two important reasons to provide a short history of human examination of coral. The first reason is that it shows that coral doesn't necessarily belong to any discipline or professional group. Coral is a tiny invertebrate, colonial, infrastructure-building organism that anchors huge oceanic ecosystems. The wide-ranging roles of the coral polyp translate into wide-ranging disciplinary, professional, and personal interest in understanding it. In the academic realm, coral researchers are geologists, paleontologists, biologists, and ecologists. In the public aquarium and hobbyist communities, people interested in whole ecosystems are as likely to keep corals as those only interested in keeping invertebrates.

The second reason exploring the history of reef research is so important is to emphasize the requirement of technological intervention when studying these organisms. As we have seen in other chapters, humans cannot observe and experiment with submarine ecosystems over the long term without visualization technologies. In the case of coral, understanding the earliest interactions with these organisms and ecosystems shows how difficult it was to work with them without building aquariums. This section will focus not on the scientific findings of the earliest research on corals but instead on the observational and experimental work done by these researchers.

During the age of oceanic exploration, coral reefs emerged as travel hazards; unmapped geological structures threatened to ground seagoing vessels. Early scientific voyages contained onboard naturalists who sought to understand the marine environment. But it was an onboard companion who also happened to be a naturalist who would push coral reefs into the spotlight. Charles Darwin was the companion to the captain of the HMS *Beagle* on a voyage around South America between 1831 and 1836. While he was not the official naturalist, Darwin performed some of this role throughout the voyage, including collecting samples. Before leaving England, Darwin's friend Charles Lyell proposed a theory of geologic change that posited gradual change in Earth's structures. Darwin joined the voyage with a geological mindset, and coral reefs offered a chance to test Lyell's theories via direct observation. Darwin observed different coral formations on the journey, and upon his return to England, he presented his findings on coral formations to the Geological Society of London. The form of research Darwin engaged in, shipboard observation, would be the primary form of coral study for almost seventy-five years.[13]

In the early twentieth century, marine-biological stations and tropical field science expanded coral research and changed the methods used from shipboard observation to in situ experimentation. The Carnegie Tropical Marine Station on the Dry Tortugas in the Florida Keys established in 1902 served as the first hub for coral experimentation in the world. The laboratory sat on an island surrounded by a natural lagoon. The two lead researchers at the station—geologist Thomas Wayland Vaughan and biologist Alfred Goldsborough Mayer, both Americans—used this lagoon to observe coral as a "natural experiment" and manipulated corals through a variety of techniques to ascertain the light, water temperature, and food requirements of a wide range of species. Mayer and Vaughan took fragments of living corals (known as *fragging*) from the lagoon and surrounding reef and planted them on terra-cotta disks; once the coral began to grow on the disk, they could be kept in the laboratory or transported throughout the lagoon and surrounding area for experiments.[14]

The availability of a variety of corals and the accessibility of a wide range of ecosystems allowed both short-term experimentation and long-term field observation. In one experiment, Vaughan placed fifteen species of coral in a lightproof box in the lagoon to ascertain how long they could survive without light. In another, Mayer brought coral into the laboratory and exposed it to varying temperatures. In this experiment he found that "the lowering of the temperature to 13.9 degree Celsius would exterminate the principal Florida reef corals, while the most important inner flat corals would survive."[15] Vaughan performed a long-term experiment by planting fragged specimens in a variety of locations within the lagoon and the shallow seas beyond the field station. He placed a variety of specimens in deeper water, on the edge of the station dock, and throughout the lagoon. Every year, Vaughan returned to his field sites to check the growth progress of these planted frags and photographically recorded those observations. This long-term experiment convinced Vaughan that different species of coral had very specific growing requirements.[16]

Similar field experiments were performed by British morphologist Charles Maurice Yonge during his work on the Great Barrier Reef in Australia in 1928. Both Yonge and Vaughan had success separating fragments of coral from larger, living pieces and planting them back into the shallow ocean, but they had limited success maintaining coral for long periods in captivity. The Dry Tortugas and the Great Barrier Reef both allowed extended coral experimentation, but neither location afforded complete control over natural variables. For instance, Vaughan had no way of tracking water temperature, food availability, effects of storm damage, and even surrounding flora and fauna in these "natural experiments." Even these well-chosen field sites could not serve the academic experimentalist fully.[17]

The Indo-Pacific Ocean and the Great Barrier Reef had long been symbols of the topics and colonial desires for and fears of the exotic Global South. The physical movement of extracted marine resources—for food, medicine, jewelry, and art—shaped European and American imaginations of these marine spaces from the medieval to the Victorian era.[18] Paintings and eventually photographs and movies reinforced these images of Edenic coral reefs populated with dangerous predators and gentle natives. In the 1930s and 1940s, photographer J. E. Williamson and filmmaker Frank Hurley both sought to render the invisible submarine world of the reef visible using the latest technologies. Williamson produced widely distributed photographs of Atlantic reefs, and he also helped develop a Bahama Islands diorama for the Field Museum in Chicago. The habitat diorama included coral, fish, and a wounded fish whose blood attracted a group of sharks to feed on it. According to Ann Elias, "The diorama was planned as an animal hunting scene to give audiences

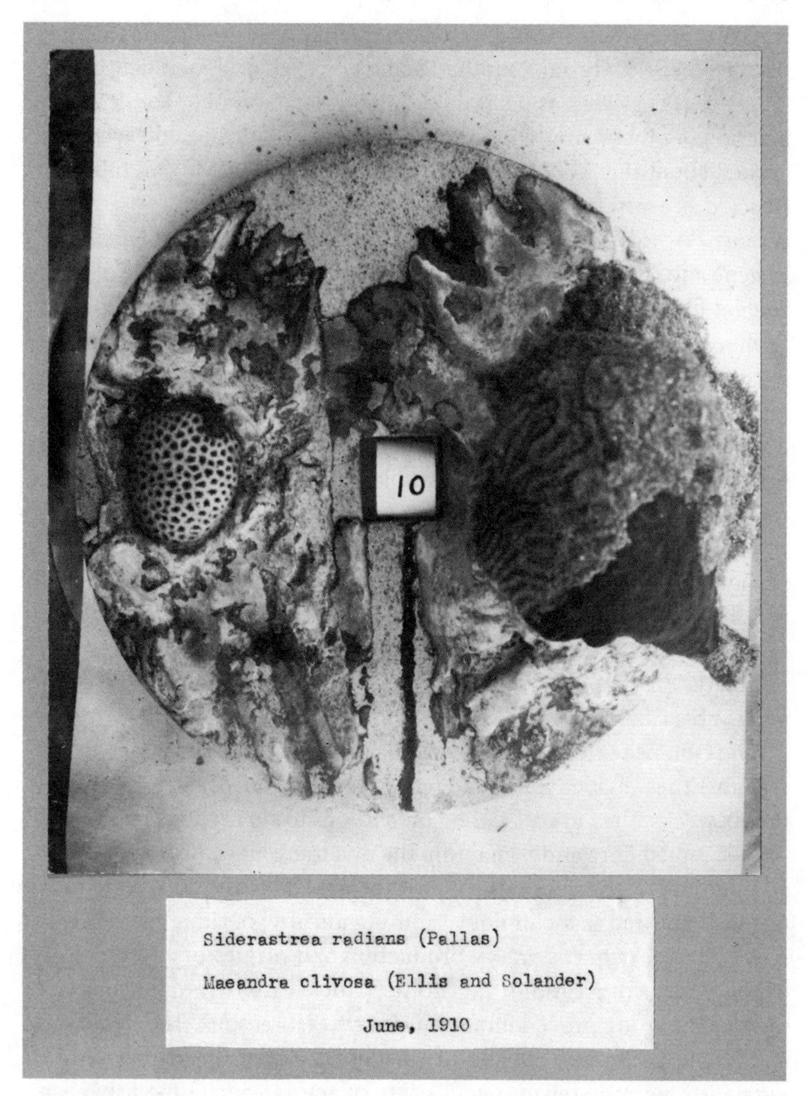

FIGURE 4.2 Tracked growth rate of corals using plates of fragments placed around the Carnegie Laboratory. Vaughan Papers, Box 8, Folder 2, "Photographs- Growth Rate of Corals Records." Smithsonian Institution Archives, Washington, DC.

in the Midwest who had never seen the ocean, or visited the tropics, the sensation of being under the sea and of witnessing the natural order of life in which sharks, the sea's top predator, give chase to prey."[19]

The perception of the reef as a beautiful but dangerous space in exotic locations was coupled with the idea that people who lived on those reefs were somehow in tune with them. Frank Hurley's films on the Great Barrier Reef and throughout the Pacific featured local men fighting and killing sharks as well as pearl divers and other local peoples. The combination of dioramas and films meant that the public became aware and excited about the reef ecosystem, but the image was one carefully constructed as exotic and other.

Post–World War II research on coral reefs by ecosystem ecologists Eugene and Howard Odum once again sought to take advantage of a convenient field site to study corals in situ. The Odum brothers belonged to a school of ecology that tried to understand the energy input and output of ecosystems to discover the role of each member in that trophic community. Put more simply, they hoped to trace the metabolism of an entire ecosystem to understand how each member of that system functioned separately and also as part of the whole. The Odums identified the coral reef as "famous for its immense concentrations of life and its complexity" and stated that in studying reefs "man can learn about optima for utilizing sunlight and raw materials."[20] The site that was chosen for their research was the Eniwetok Atoll in the Pacific. The United States nuclear testing in the area gave the Odums an opportunity to spend time at the marine station on Eniwetok and to possibly explore the ecosystem's response to nuclear blasts.

The Odums performed a six-week survey of the island's coral reef ecosystems. They divided the inland reef system into quadrants. Throughout the six weeks, they used a variety of methods to survey those quadrants. They sampled flora and fauna from the quadrants to estimate the variety and number of species on the reef. Drawings of ecosystems were made using diving mask technology, and chemical experiments were performed to ascertain "primary production and respiratory metabolism" of the reef.[21] In addition, the brothers took extensive light measurements. The study proved important for several reasons. The first is that the Odums confirmed the idea of the reef ecosystem as existing within a nutrient-poor area reliant on a variety of actors—including fish, algae, coral, and water movement—to function. The second is that this study jump-started a new interest in studying coral reefs throughout the world.

This brief summary of reef research shows that a wide range of people had been thinking about and researching these spaces for a long time, but the next phase in studying them would widen the network of researchers and technologies and bring reefs indoors. The development of the reef tank made it possible to break reefs down into their individual components and study them for the long term, and it prompted deepening questions about what a reef actually is and whether we can ever fully understand it.

EARLY REEF TANKS: BOUND BY LOCATION

Reef tanks arose in the Indo-Pacific because of a combination of traditional knowledge, location, and species availability. The Pacific Ocean contains both hardier and more varied species of coral than the Atlantic. These corals are extremely colorful and live in shallow, clear water. This contrasts to Atlantic corals that have a tendency to be less flashy and colorful and often share space in more turbid waters with algae and seaweed. The plentiful, beautiful, and hardy nature of Pacific corals made them an optimal candidate for successful captivity. The availability of coral—combined with increased military, scientific, and hobbyist activity in the area after World War II from both residents and European immigrants—produced the first reef aquarists and some of the earliest theories about captive reef building. Advances made in the Pacific during this period spread throughout the world through hobbyist journals, jump-starting the development of reef tank technology in areas that were less hospitable.

René Catala and Ida Catala-Stucki[22] built the first successful reef tank at the Noumea Aquarium in the French territory of New Caledonia in 1956. René Catala grew up in the Vosges region of France and received his doctorate in natural history from the University of Paris. In 1945 he was sent to New Caledonia by the French Colonial Research Office to participate in the founding of the French Oceania Institute (Institut français d'Océanie [IFO]), which became the Overseas Office for Science and Technological Research (Office de la recherche scientifique et technique d'outre-mer [ORSTOM]) in 1964 and the Institute for Research and Development (Institut de la recherche pour le développement [IRD]) in 1998. While his earliest work was done on moths in New Caledonia, René became entranced by the corals populating the shallow reefs just off the coast. René and his wife Ida began building the aquarium in 1946, personally financing the endeavor with help from the Noumea city government to finish the project. It was in this aquarium, now known as the Lagoon Aquarium, that the first successful reef tanks were built.[23]

The location of the aquarium made it possible to collect and transport delicate corals to the aquarium without damaging them. The transportation of corals from a reef to an aquarium was one of the first major hurdles in setting up a reef tank. The short distance from the aquarium to the reef, ten miles, made it possible for the Catala-Stuckis to maintain coral health during transportation. But they still had to take special precautions to avoid damaging their specimens. Ida Catala-Stucki, referred to simply as "Stucki" in René's publications, collected corals with a team of divers, each using the newly popularized scuba.[24] Each diver collected with "his own basket, a little wooden frame with chicken wire sides, where selected specimens will be deposited." After each diver returned to the ship, their baskets were carefully raised from the ocean floor, and the corals—"completely retracted and crumpled" and covered with mucus "as a defense mechanism"—were loaded with the "greatest care" so that "neither pitch nor roll can make them knock against each other." The full collecting boats proceeded back to the aquarium at a sedate pace. Upon arrival, René placed the corals in a sandy bottomed holding container in low light with constantly running seawater to acclimate them.

The collected corals were eventually placed on display in the aquarium in the first reef tanks. The Catala-Stuckis placed multiple coral species together in tanks and included invertebrates and fish. Because of their location, they had a fairly simple tank design. The ten tanks were made of glass with a sandy bottom. Unfiltered seawater from a reef near the aquarium was pumped continuously through the tank. Light was provided from skylights in the aquarium. The indoor tanks were extensions of the outdoor reefs, using as many unchanged elements as possible to maintain an environment in which the coral could thrive. But unlike the reefs, these tanks made the reef visible for continuous experimentation and observation. The Catala-Stuckis emphasized the importance of visualization in these tanks and built specialized reef tanks outside on ball bearings so they could be rotated "to take still and moving pictures."[25]

These tanks allowed the Catala-Stuckis to make new observations about corals. For instance, René begins his book detailing the way that coral reproduces. But he also states that

> Observations made recently in the Aquarium have shown that reproduction can also take place in other ways. A colony of *Goniopora* comprising about 200 polyps was seen to release the male cells about every 10 minutes over a period of an hour, while a nearby colony of the same species was releasing eggs. Fertilization may thus take place in open water.[26]

René also found that many of his corals bioluminesced, leading to groundbreaking images in his book *Carnival under the Sea* and an invitation to develop an aquarium show in Antwerp. The Catala-Stuckis eventually became the first aquarists to display live corals in Europe, detailing their method for transporting them via plane in *Carnival under the Sea* as well. Those corals survived a mere two months in Antwerp, many dying within a few weeks of arrival. This episode shows that much of the success of the reef tanks at Noumea can be attributed to the minimal distance the corals needed to travel to the tank and the use of natural sunlight and unfiltered seawater in the tank system.

In 1961 hobbyist and aquarist Lee Chin Eng, an ethnically Chinese man living in Indonesia, used a similar system to the Catala-Stuckis, which he labeled "Nature's system" of reef keeping. Unfortunately, there is little information about Lee, although he captions an image as Lee of Prinsen Park Aquarium and a fellow aquarist, Mr. Saunan, from the Pasar Ikan Aquarium. Lee points to his friend Mr. Tan Soen Hway of Bangu-wangi as "a pioneer of discovering Nature's system of keeping fish in 1955."[27] The recognition that Lee was passing along craft knowledge developed and recognized throughout Indonesia means that the history of reef tanks might be much deeper in Indonesia than written information suggests. Lee, Mr. Tan, and Mr. Saunan all appear in a picture showing them preparing to snorkel to collect fish and coral, suggesting that they had a particularly short transfer time from sea to tank.

Similar to the Catala-Stuckis, Lee Chin Eng developed a special method for protecting coral during and after collection. In the images, the collectors wear masks. The coral, rock, and fish that Lee and his colleagues collected were immediately placed in a basket. That basket was held up in the water via an inflated inner tube and the basket fit inside the inner tube with the bottom hanging into the water. This vessel was designed to contain the organisms collected while keeping them wet. The inner tube system allowed the basket to float lightly on the water and kept the coral from being injured as it was transferred to shore, and eventually, the reef tank.[28] But while Lee had much to teach regarding his collection techniques, the hobbyist community was more interested in his practice of maintaining his reef tank.

In the February 1961 issue of *Tropical Fish Hobbyist*, Lee reported on his system for maintaining a "balanced" reef tank. Lee describes his process thus:

> Follow Nature as closely as possible; all can be done except wave and current in the aquarium, but this could be duplicated with an airstone. Set up a non-toxic aquarium, collect and pour in real sea water of a density approximating 1.022. Don't filter the water but keep the aerator running in order to keep the plankton living. Select nicely shaped corals from the sea that contain microscopic plants and sea-weeds, arrange the corals all over the bottom and with many hiding-places in the back. Never boil or bleach the corals; if you do, you kill all the necessary plants, worms and other marine animals which live among these corals and which are the very important things that help balance the aquarium. If you put in the right corals, within a day or two the water will become crystal clear, after which you may put in anemones for your Clownfishes. Depending on the size, your aquarium can be decorated with: small

FIGURE 4.3 Lee Chin Eng in front of an early reef tank. Lee was a hobbyist, but also built tanks for other hobbyists. This is an image of him, probably from the late 1960s, in front of a tank that he built for another hobbyist. It circulates online and can be found at https://www.nano-reef.com/forums/topic/412435-a-tribute-to-lee-chin-eng/.

> multicolored anemones, colored mushroom anemones, tube worms and live corals (brain corals, mushroom corals, etc.) and fishes, at least twice as many as your text-books recommend.[29]

Lee fails to describe his lighting setup, but based on the photos he included with his article, it appears that he used a combination of artificial and natural light.

His article caused a stir in the aquarium-keeping community, many of whom felt that there was no way to successfully keep coral in home aquariums and that a balanced system, represented by a reef tank, was impossible. The editor of *Tropical Fish Hobbyist*, Herbert Axelrod (see chap. 2), stated in a note at the end of the article that he "believed Eng implicitly" but couldn't fathom how any of his readers would get their hands on living coral. The addition of the editor's note, assuring the audience that the editor believed Lee's assertions but questioning the ability to transfer the techniques presented, shows how new the idea of keeping reef tanks was during this period. But the circulation of Lee's article and René Catala's *Carnival under the Sea* began a rapid development of tank designs and techniques dedicated to building and operating reef tanks.

Three Technological Systems of Reef Building

The reef tanks built by the Catala-Stuckis and Lee functioned to maintain ecosystems, but they were largely location bound. The delicacy of coral collection and movement along with the requirements of unfiltered, running water; fresh coral and sand; and direct sunlight kept the practice of keeping reef tanks isolated to tropical regions. But the practice did spread. Paul Jokiel at the Hawaii Institute of Marine Biology built some of the earliest experimental microcosms to study the effect of temperature on climate health. Jokiel's systems used natural sunlight and fresh water run through the tanks from the surrounding ocean. All the corals were collected nearby and placed in tanks with access to full sunlight and heavily aerated natural seawater. Jokiel's work concentrated on the importance of UV exposure to reef death and the effect of both increased and decreased temperatures in reef health.[30] Another study conducted with Stephen L. Coles showed how useful microcosms are for isolating essential elements. Coles and Jokiel used tanks to control three elements: temperature, salinity of water, and sunlight exposure. By using microcosms and changing these variables, the authors found that low salinity and high light exposure both exacerbate temperature-based damage in corals.[31] In another experiment with Coles and Clark R. Lewis, Jokiel used tanks in Hawaii and on Enewetak (where the Odums worked) to compare the survivability of corals at different temperatures; the study found that Hawaiian corals are more sensitive to temperature variations than those corals on Enewetak.[32]

At the same time, Bruce Carlson, the head of the Waikiki Aquarium in Hawaii, successfully kept a variety of coral species in tanks throughout the 1970s. These reef tanks were very close to the point of collection; in some cases, the aquariums were right next to the water in Hawaii. The setup was like those described in the previous section, with unfiltered water and natural sunlight. Both Carlson and Jokiel's work shows the way various communities worked to keep coral in captivity; both tanks required natural light and seawater, and corals could not be moved far from the collection site. Carlson's work also highlights not just the spread of reef keeping but the entrance of reefs into mainstream public aquarium exhibits. This is an important step, because as the process of keeping these corals became known and discussed, other aquarists, hobbyists, and scientists sought to develop a practice that would allow them to exhibit these ecosystems inland.[33]

Three systems for building and maintaining reef tanks were developed from the 1970s through the early 1990s. During this period live coral became a mainstream product in the pet trade, and public aquariums began collecting and displaying coral regularly, decoupling the keeping of these organisms from the natural space from which they came. All three of the tanks developed built on the idea of maintaining a system that closely simulates natural ecosystems. However, what variables were recognized as natural and how closely one wanted to hew to those natural conditions fractured the reef tank community.

BERLIN SYSTEM

Peter Wilkens introduced the first of the three systems of reef tank building in 1973. Wilkens was a Berlin-based hobbyist (a professional water analyst by day) who belonged to the Berlin Marine Aquarium Association. Another member of this association, Dietrich Stüber, is credited as the first person to maintain stony corals in captivity for extended periods. The "Berlin System," as it is called, is a system of building and maintaining reef tanks that did not necessarily originate with Wilkens but was popularized by him in his book *The Saltwater Aquarium for Marine Invertebrates*. This book outlined the techniques developed by the Berlin Aquarium Association at their meetings that led to success in reef building.

The Berlin Method simulates natural ecosystems with the help of careful filtration and the addition of trace elements. Unlike the earlier "natural system" of Lee, which relied as much on the environment in which the tank was built as the act of tank maintenance, Wilkens developed his method so tanks could be maintained in northern climates. Wilkens suggested the use of "living rock,"[34] which is rock that is taken directly from the water source and placed into the aquarium without sterilization. He admitted that "the reason for this is not completely clear," but he believed that "from my observations over the last few years, living rocks help balance the small biotope and stabilize the ecosystem in our tanks."[35] In addition to the organisms brought into the tank via living rock, Wilkens also emphasized the importance of maintaining a healthy, but balanced, algae community in the tank. Unlike other systems, which sought to minimize algae in the tank for aesthetic purposes, Wilkens emphasized that in reef tanks "the algae play an important role in providing living space for the micro-fauna. They offer protection and food to uni-cellular structures up to the small crabs."[36] Because algae were "very valuable" as food and living space in the tank, "we therefore readily put up with the occasional opacity of the water."[37] The attention to the needs of the ecosystem over the aesthetic desires of the aquarist shows the way that the aquarium community measured success in these cases: through the ability to maintain balance and biodiversity for long periods.

Unlike earlier iterations of systems, Wilkens used filtered, prepared seawater. The mixed seawater was then added to the system along with the organisms. Instead of eschewing filtration, as Lee suggested, Wilkens used a protein skimmer to maintain proper nitrogen levels. Protein skimmers remove excess nitrogen from animal waste and food by pushing them into a chamber filled with little bubbles. The protein collects on the bubbles and is taken to the top of the device, where it is removed by skimming from the top. As one would expect, without natural sunlight, the Berlin Method called for very strong illumination with halide lighting and attention to night and day cycles. Corals should be exposed to light in a seasonal pattern, paying close attention to the native environment where they were collected to determine the amount of sunlight required at a given time.

The most lasting contribution of the Berlin System is the method of adding trace elements to achieve coral health and growth. Because the Berlin System used filtered water, Wilkens suggested adding certain trace elements into the water to mimic elements found naturally in wild reef environments. Wilkens developed a product called Combi-San, still in use today by reef keepers. The product contains trace elements that can be removed by filtering water through protein skimmers, including calcium. Calcium is important to place back into a reef system because stony corals build hard skeletons by pulling calcium from the water. Another additive Wilkens introduced to the reef-building community was kalkwasser; this is a calcium additive that also helps with stony coral building. These additives are required in the Berlin System because protein skimming removes trace elements. This system worked well for maintaining corals, but it required extensive technology, such as protein skimmers, and the availability of additives, many of which were already patented and being sold via aquarium supply companies.

With the Berlin System, Wilkens was able to expand the types of corals that could be maintained in northern climates. In a 1974 feature in the hobbyist magazine *Marine Aquarist*, Wilkens states that

> among the large class Anthozoa . . . we find the most conspicuous and durable candidates for a marine aquarium, such as anemones (Actinia-

> ria), tube anemones (Ceriantharis), and the small encrusting nemones (Zoantharia). Other orders within the class Anthozoa, such as the soft corals (Alovonaris), sea feathers (Pennatularia), or true stone corals (Scleractinia or Madreporaria) are more difficult to maintain because of their specialized environmental requirements. But with the technical equipment available today, even these delicate "flower animals" can be kept for long periods of time in the home aquarium.[38]

In the article, Wilkens describes the process of introducing and observing a gorgonian coral (sea whip) in his "mini-reef." He describes himself as "quite occupied" with the whip for the first few weeks and spending time "feeding it minute particles of finely lacerated deep frozen or freeze-dried shrimp, crab meat, mussel flesh, and freshwater copepods, as well as *Artemia* nauplii."[39] The whip grew rapidly, and six months later, Wilkens added a sea fan to this reef tank as well.

The expansion of new coral maintenance capabilities also expanded the combinations of organisms in these mini reefs. Wilkens states that the horny corals are too sensitive and brittle to maintain in the same tank with fish. Instead,

> The upper portion of the "Mini-Reef" is populated with soft corals and encrusting anemones, as well as a few true stone corals (all require a great deal of light). As scavengers I have three small brittle stars. These are as efficient as shrimps or hermit crabs in finding every food particle which becomes lodged among the coral and decorations. A file shell and a large reef clam (*Tridacna*) are also at home here.[40]

The requirements to maintain this system seem relatively small to Wilkens: a protein skimmer and a power filter were "absolutely necessary to imitate the aquatic environment of these delicate invertebrates," but feeding was simple—freeze-dried foods, flake food, or small crustaceans could be offered to every member of the community.[41]

The Berlin hobbyist group became known for their "mini reefs," and they spread both craft and specimens throughout the world. When Lee Chin Eng introduced the concept of live rock to hobbyists, very few aquarists were using this method in their tanks. But he paired with Merrill Cohen, a commercial hobbyist in Baltimore, Maryland, to export live rock from Indonesia into European and American markets.[42] Live rock was the cornerstone of the Berlin Method, and its use brought with it surprises. In the late 1980s Dietrich Stüber discovered a small staghorn coral growing on a piece of live rock in his tank. This hitchhiker, photographed by Svein Fosså in 1985, became the first instance of a small polyp stony

FIGURE 4.4 Dietrich Stüber's aquarium in West Berlin. The Berlin hobbyist group had great success maintaining stony corals in their "mini-reefs" beginning in the 1980s. Svein Fosså and Alf Jacob Nilsen documented this community of aquarists and their successes with a series of photographs in 1985. Svein A. Fosså and Alf Jacob Nilsen.

coral growing in captivity. Stüber passed frags of the coral throughout the Berlin network, and it spread from there into the larger hobbyist and public aquarist communities. Today, Stüber's *Acropora* is maintained throughout the world.[43] Other specimens, such as Grube's gorgonian, also circulated widely. On a website for the commercial group Oceans, Reefs, and Aquariums (ORA), the story of the gorgonian's movements are recounted by Julian Sprung. According to Sprung, the initial sample

> was originally collected in the Philippines around 1990 by Klaus and Rosalia Grube. From their aquarium in Berlin it has spread all over Europe. So many people have it but don't know where it came from. I got my colony from the Grubes and grew it for several years into a big bush. I gave a cutting to Peter Wilkens, who took it back to Switzerland and grew it at Wiwi Aquaria. He gave a cutting to Jean Jacques Eckert in France, who still cultivates it in his aquarium. I lost my original colony

> suddenly about 12 years ago, but was able to get a new cutting from Jean Jacques Eckert during a visit with him for a Récif France conference. I've been careful to maintain it now in several tanks. For me this coral is always a reminder of my visit to Berlin where I met Dietrich Stüber and the many wonderful hobbyists there, including Klaus and Rose Grube.[44]

We can see that the Berlin community, and Wilkens's writings, spread both the craft of reef keeping and the materials used by that community throughout the world.

Wilkens experimented and published his findings with tanks throughout his life. In a 1975 article in *Marine Aquarist*, he reported on an experimental tank he began at a marine-biological station in 1969. The tank was built to provide researchers opportunities to observe delicate marine flora and fauna. The article contained a list of every animal he had kept in the tank, its life span in the tank, and the conditions of the tank throughout that period. This extensive information and the continued reporting on these projects show how important it was for him to advance his craft and share his findings with both the wider hobbyist and academic communities.[45] But Wilkens was only one of several aquarists working on perfecting the reef tank.

JAUBERT OR MONACO SYSTEM

Jean Jaubert introduced the second reef tank system, known as the Jaubert Method or Monaco System, at the Second International Aquarium Conference in 1988. Jaubert was born in Oran, Algeria, in 1941. He received a PhD in biological oceanography from the University of Marseille in 1971 and a Doctor of Science from the University of Nice in 1987. He worked throughout the 1970s as a marine researcher and lecturer at the University of Nice. Jaubert was a hobbyist from a young age (nine years old) and kept aquariums throughout his life. He successfully kept both soft and stony corals alive in captivity in the 1960s. In 1973 Jacques Cousteau asked Jaubert to set up a reef tank at the Musée Océanographique de Monaco; Jaubert collected fifteen species of coral from the Mediterranean Sea and built a tank. However, he experienced coral loss from algal blooms soon after introducing fish to the tank. The same losses occurred in his laboratory tanks at the University of Nice. Jaubert, having not read Wilkens's work about using protein skimmers, sought to decrease the nitrogen in the tank (the cause of algae growth in the system) another way. After fifteen years of trial and error, Jaubert presented his new tank design in 1988.

To reduce the nitrogen in the tank, Jaubert built a system that trapped a layer of water under the sediment; he called this space a *plenum*. When building the tank, a layer of hard plastic grating was applied to the bottom of the tank. Then, a layer of coarse sand was added, followed by a thick layer of sediment, and another layer of finer sand. This thick sediment traps a layer of water underneath the sediment. Organisms burrow into the fine sand and live in the thicker layer of sediment; these organisms eat the nitrogen-rich food waste from the tank and deposit it, via their own waste, farther into the layer of sediment, where it eventually filters into the plenum. The plenum traps the nitrogen in the bottom layers of the system and balances the chemistry of the aquarium water without using protein skimming. Jaubert stated that in the three years that the aquarium had been running, his corals thrived, and he did change the water.[46]

Jaubert makes it clear that he considers himself both a hobbyist and a research aquarist in his work with reef tanks. In his original publication about his system, he states that the result of his tinkering was a physico-chemical balance of a complex ecosystem. He built a "miniature-reef" at the University of Nice meant for observation. But the tank had multiple uses: "this system has direct application ranging from scientific research to public and home aquaria of many sizes."[47] In later publications about his system, Jaubert again asserts his hobbyist status while also calling attention to the scientific applications for his "mesocosm": "This method enabled me to set up a 40,000 L reef tank in the [*sic*] Monaco's Oceanographic Museum and to resume research in this famous organization by establishing the European oceanographic center (affiliated to the Council of Europe) where the ecology and physiology of cultivated corals was investigated using the most advanced scientific techniques."[48]

He states that continued use of the microcosm has resulted in research findings, such as that elevated temperature and carbon dioxide partial pressure (pCO_2) have a synergetic action, which inhibits calcification.[49] According to Jaubert, "This result may explain why reefs, which are exposed to warming temperatures and surface water acidification associated to rising concentrations of atmospheric CO_2, degrade worldwide in remote areas where they are not directly impacted by human activities."[50] Jaubert continues to be a speaker at academic and hobbyist conferences, but his work was not the last mesocosm in this period.

While Jaubert had not heard of Wilkens's system or Lee Chin Eng's work, he had read the work of another reef builder named Walter Adey. Jaubert chose to build a closed-circuit system, but in his 1988 paper he called attention to Adey's biological filtering system. While Adey had been working on his system for over fifteen years, he and his wife and research partner Karen Loveland would not write a book introducing his methods to the larger hobbyist and aquarist community until 1993.

ADEY SYSTEM

Walter Adey and Karen Loveland developed the last system for maintaining reef tanks during this period. Adey was a researcher and curator at the Smithsonian National Museum of Natural History (NMNH) working on geology, paleontology, and botany. His work focused on the geologic development of coral reefs. Adey's research led him to question the best way to study existing coral communities to understand the life cycle of coral reefs. His wife, Karen Loveland, was a filmmaker in the Smithsonian's Office of Communications. Studies such as those performed by Odum and Odum showed that long-term reef observation was difficult. Experimental research, involving the manipulation of variables to observe outcomes, was nearly impossible on an open-water reef. Adey and Loveland felt that the best way to study coral reefs was to build a working model ecosystem in the laboratory.[51]

In 1978 Adey and Loveland built their first reef tank in the basement of the NMNH and followed this with a larger tank for display in 1982. Adey and Loveland collected their corals from around the Caribbean and tried to collect as many species from the same locations as possible so that the tank would reflect the natural environment. Adey's tank design followed the same idea as those previously discussed: he wanted to recreate an actual ecosystem indoors. As an academic biologist, it might be suggested that Adey would want to build a simplified model to help researchers, including himself, easily identify and manipulate variables within the system during scientific research. Instead, Adey wanted to recreate a reef without trying to track all the variables before building it. According to Adey, it was this very aspect of the "black box" tank that would make the system so useful for studying coral reefs. In a promotional video meant to air in the NMNH next to the tank, Loveland recorded Adey talking about the tank. He states that because the variables of the tank cannot be controlled, the tank gives "results [that] can be trusted" to tell you something about the natural systems the tank models. Adey felt that including complexity, even complexity that the tank maintainers didn't understand fully, was important to really studying a system. Therefore, Adey had to "figure out how to open up our box and let the open ocean in."[52] To achieve this type of natural experimental model indoors, Adey introduced two previously unseen elements to his tank design.

The first, and possibly the most controversial aspect of Adey's tank design, was his development of an Algal Turf Scrubber (ATS). According to Adey, the use of more common filtration systems did not mimic the natural filtration taking place on coral reefs. When diving, Adey noticed a layer of algae around most coral reefs. Based on his ecological knowledge and diving, Adey believed that algae actively worked as a filter and oxygenator on the reef. Basically, algae grows as it absorbs excess nitrogen from dead organisms and waste product from reef dwellers. In return it produces oxygen for the reef, "smoothing out the differences between day and night in small model ecosystems or between seasons in large mesocosms."[53] Adey believed that algae were an integral part of the reef ecosystem and sought to include them in his tank design. Traditional thinking in aquarium communities is that algae are generally bad for a system because they can grow out of control and destroy the coral and choke the entire tank. Algal infestations are by far the most consistent problem for those aquarists seeking advice in the columns of hobbyist magazines. Adey sought to maintain a balanced tank with algae and produced the ATS to do this.

The ATS works as follows. A woven mat of either plastic or cloth is seeded with a variety of alga species. It is as easy as taking algae found on rocks or in aquatic systems and spreading them on the mat. The mat is then hooked into the filtration loop by placing it on a long, flat board that is then placed in the tank filtration cycle in a position to have water collect onto it. Water is pumped from the tank onto the ATS mat, and once the board becomes overfull, it tips the water back into the tank (called a dump-bucket type ATS; US Pat. No. 4,966,096). As the algae grow, they clear impurities from the water (the algae feed on them) as that water passes over the ATS. As the water fills in the ATS, it is reoxygenated by the output of the algae. The dumping of the water back into the tank from the ATS also helps mimic wave action and adds more oxygen into the system. Adey states that "the key to ecosystem management is stability achieved by locking nutrients up in biomass rather than by using bacterial filtration to rapidly reduce all nonliving organics and organism excretions to freely available elements and ions."[54]

According to Adey's original designs, the ATS was the only filtration required to maintain a balanced reef tank. However, other hobbyists and aquarists were not completely sold on Adey's ATS system. Much of this disagreement can be chalked up to differing expectations and requirements of their tank systems. The ATS created a water quality that many have described as yellowish or somewhat opaque. For many hobbyists and public aquarists, the pursuit of the most natural system came second to aesthetics. If a tank functioned but was unattractive, it would not work for these groups. In addition, many suggested that the ATS was only adequate for the thriving of nonstony reef corals. Hobbyists hoping to grow reef-building corals felt that the ATS did not adequately create water qualities needed to do so and so used the Berlin System instead.

The second major contribution Adey made to reef tank design was the use of refugia. In ecology, a refugium is defined as a protected habitat for species to survive in unfavorable conditions. Examples in terrestrial population biology include areas that harbored small numbers of individuals of a species during the last ice age; the refugia were created by glaciation, and those isolated areas allowed species to become more robust before the glacier-made impediment was removed and predation resumed.[55] According to Adey, refugia are needed in reef tanks because of the size of the simulation. One concession to the idea of "naturalness" in his reef tank was the acknowledgment that most species in the tank required more space to escape predation and to procreate. Because a reef tank, by definition, contains an ecosystem filled with both predators and prey, this lack of space could result in an imbalance favoring those animals higher on the food chain. The Jaubert and Berlin systems had what might be considered refugia, with the addition of rock spaces and sand that could be colonized by smaller organisms until they enter the water column. But Adey's system called for refugia separate from the main tank.

> Thus, after economic constraints have determined the maximum size of the model, then refugia are added with the intention of circumventing obvious population constraints of selected elements of the community. Most of our refugia have been directed toward freeing attached fleshy algae, soft-bottom invertebrate populations, and plankton from severe predation by fish and larger invertebrates (particularly crabs and lobsters).[56]

Adey's refugia sit above or next to the larger reef tank and are hooked into the water flow system so that as animal populations become mature and capable of defending themselves in the larger ecosystem, they do not need to be acclimated a second time but can be added directly to the main tank. In some of his designs, including the tank at the NMNH, a sea grass refugium included a population of invertebrates that served as food for the fish in the tank. As they matured, they entered the water column and drifted into the larger tank, where they were preyed on by those vertebrates.

The use of refugia was more widely accepted than the ATS by the hobbyist and public aquarium communities, probably because it is a common practice in breeding systems (see chap. 5) and because it amended a problem without requiring a large amount of new technology. When a reader asked Julian Sprung his opinion of Adey's system in the "Reef Notes" column of the *Freshwater and Marine Aquarium*, he responded that the ATS was not necessarily useful for many types of tanks, but that

FIGURE 4.5 Reef tank by Walter Adey. You can see both coral and a sea urchin in this early image of the tank (1979). Smithsonian Institution Archives. SIA2009-2204 Farrar, Richard.

> Dr. Adey's idea of using external reservoirs is an ingenious way to accomplish two tasks. Such reservoirs can be used to settle out detritus, thereby eliminating the need for a mechanical filter, and one can also use these refugia to cultivate various microcrustaceans which can serve as food for the system. Grass shrimps, mysis, peppermint shrimps, and others may also be maintained in these refugia where the fish won't eat them, and their offspring will drift back into the main display where they are rapidly consumed by fish and invertebrates.[57]

The biggest arguments for or against Adey's system concerned not an addition to his tanks but an omission. Adey believed that the use of the ATS balanced the system so well that the addition of other elements, including calcium for stony coral growth, was unneeded. Adding trace elements and balancing the system via filtration allowed users to create a reef tank that could be maintained without spending years finding the correct balance between all the organisms. For hobbyists and public aquarists, this was important to prevent loss but also to develop a tank that contained the organisms that they wanted, not just the ones that would work within that grouping. Adey's tank system wasn't for the collector or displayer but was more geared toward the long-term marine tinkerer interested not just in display but in the challenge of creating a "more natural" ecosystem. Aside from these two elements, Adey's system mirrored those of the Berlin and Jaubert systems. It used illumination via artificial light and attention to timing with lighting, the use of live rock and substrate, and the balancing of flora and fauna.

Adey's tank never became a well-used scientific tool, but its life cycle demonstrates the way that reef tanks and tank keeping have shifted since the 1990s. The aesthetics of Adey's tank did not lend itself to visitor satisfaction at the NMNH, and it was eventually moved to Fort Pierce, Florida. Bill Hoffman began working with the tank at the NMNH and moved to Fort Pierce when it did, setting up the tank with the same corals and working to calibrate the system for optimal ecosystem health instead of continuing the idea of modeling the natural state perfectly. To do this, Hoffman continued using the ATS system but added additional filters and began supplementing the water with elements to enhance coral health and growth. He continues to use live rock and live water, both taken from the waters right off Fort Pierce. As UV lights became more popular and affordable, they were added (in addition to the traditional halide lighting). After a series of growing pains and disease epidemics, the tank is stable, and corals, especially the endangered staghorn coral, thrive. The current iteration of Adey's tank uses a combination of techniques and technologies that combine the Jaubert, Berlin, and Adey methods to maintain the system.[58]

Linking Oceans and Oceans under Glass

At the same time that debates about tank design raged, reef tank aquarists openly discussed questions about how much tanks could tell us about natural reef systems. While the earliest tanks were practically *on* reefs, the farther these tanks moved inland, the more debate arose in the aquarist community about how useful they were as simulations of reef ecosystems. While the entire aquarist network agreed that it should be possible to build tanks to test scientific theories about coral reefs, they disagreed on the structure that this knowledge would take and who could produce it.

One reason that researchers felt it was difficult to produce knowledge about the ocean with reef tanks was that these tanks did not adequately allow for the control of all variables in an experiment. The very definition of a reef tank is a system with a variety of species all living in a "balanced" ecosystem. Over the years, Adey has worked to get academic researchers to use reef tanks. He designed the reef systems for Biosphere 2 with the hope that these systems would become useful in studying connections between the ocean and terrestrial forms. In addition, he has coauthored academic papers in the hopes that more researchers will utilize these systems.

However, even those papers contain muted enthusiasm and trust in these systems for knowledge creation. In a 1996 paper in *Ecological Engineering*, Adey uses the example of his five-hundred-liter aquarium that he and Loveland have maintained in their home since 1988. The tank was built with a mix of Indian and Caribbean Ocean species and successfully spawned and settled stony corals. The successful coral propagation and long-term maintenance of this system shows expertise in the hobbyist community, but the coauthors of this academic papers highlight that "such mixed-faunal ecosystems are not condoned by the authors due to the potential for unnatural species interactions that are ecologically difficult to assess. However, this unit with its reproductive successes, provides a valuable reference point in terms of comparison despite its shortcomings."[59] In this way, it seems that academic researchers are not quite ready to accept Adey's "black box" as a useful scientific tool.

The distrust of reef tanks as systems capable of producing formal knowledge about the ocean can be seen primarily as a debate about the methods of proof required by academic researchers to accept something as knowledge. In the reef tank community, knowledge takes the form of successful tinkering: if something works and the tank thrives, this tells us something about the needs of corals both in the tank and in the wider world. But the lack of scientific rigor is not lost on the community. Jean Jaubert stated in a 2008 paper,

> The plenum is the void space situated under the plastic screen that supports the coarse coral sand on the aquarium floor. The plenum contains a layer of hypoxic water. Its role is not substantiated by scientific data. I could never find enough time and money to carry out the experiments needed to compare the denitrification yield of aquariums functioning with and without plenums. Indeed, to comply with the statistical constraints of scientific experiments, I should have set up a minimum of 3 aquariums with a plenum and 3 aquariums without a plenum. Nevertheless, I presume that the plenum facilitates pore-water circulation through the sediment in aquaria within which stirring is moderate and that pore-water circulation plays a key role.[60]

Adey and Jaubert, both with advanced degrees and working within the academic community, understood the limits of their systems with regard to academic knowledge production but believed them to be important tools for understanding the ocean regardless.

The lack of strict variable control and comparative experimentation in reef tanks has pushed some to suggest that aquarists should seek to hew more closely to the academic system to produce knowledge that is taken seriously by that community. Richard Ross, an aquarist at the Steinhart Aquarium and writer of the column Skeptical Reefkeeping in *Reefs Magazine* suggests that hobbyists should seek to become hobbyist-scientists. Ross suggests that aquarists have too commonly relied on knowing that something works instead of how it works. This leads to anecdotal evidence and generally unreliable information in the hobby. In a 2016 article titled "Everyone Can Do Science," Ross urged his readers to be more methodical in their craft development. According to Ross,

> In reef keeping, there are a million products and techniques, each claiming to be a necessary ingredient for a successful reef tank. The problem is that many of these claims have little, if any, evidence to support them—so how are we supposed to know which ingredient, product or method is useful and which is bunk? Well, we can bellyache that someone else should figure it out and let us know, or we can get up off our collective butts and start producing evidence ourselves by doing some simple experiments.[61]

In the bulk of the article, Ross outlines a set of simple control experiments for finding out whether new tank additives work.

Ross is not the only aquarist to call for a turn toward more rigor in reef tank maintenance and hypothesis testing. Bruce Carlson has stated that reef tanks are in the "'natural history' phase of observation, description, and hypothesis development," and he points to recent research with coral tanks (including Adey's Biosphere 2 tank) that have used tanks in a rigorous way. He champions the use of the reef tank as a model of the wild reef and acknowledges its limitations.

> The fact that an aquarium is not, and probably never will be, an exact simulation of nature, should not detract from its potential power to detect biological responses to extreme environmental changes. In this respect, aquariums are analogous to mathematical models which also are not exact simulations of nature but nonetheless can offer predictive power and insight into the nature of biological systems.[62]

But unlike Ross, who urges hobbyists specifically to become academic researchers, Carlson acknowledges that reef tanks could be used for hypothesis testing but recognizes that the usefulness of the knowledge provided by hobbyists and reef aquarists does not need adhere to academic science principles. According to Carlson, "Corals do not survive equally well in all aquarium systems and the reasons why may be instructive in understanding how coral reefs function in nature." Carlson acknowledges the difficulty of maintaining corals in captivity and suggests that paying attention to the differences in captive reef systems currently in operation can *at least* generate questions and hypotheses for further marine study. In the paper, he highlights all the information about water chemistry and food requirements that captive systems have uncovered. It is clear that work with reef tanks has resulted in knowledge about the ocean's coral reefs; unfortunately, it still seems that these tanks have not been fully accepted as scientific tools.

Craft Work Goes to Work

Tracing the history of reef tank craft allows us to see the articulation of debates about the role of tanks in marine knowledge production. The earliest iterations of reefs under glass were deemed "natural," but consistent changes to that system split aquarists into factions seeking to find the essential elements of wild reefs through tinkering with water chemistry, alga communities, and the physical structure of the substrate. As reefers gained confidence in their craft, reef tanks spread throughout a network of users hoping to use them to produce knowledge about complex reef ecosystems with laboratory simulations. But the debates in the wider reefer network were representative of academic concerns about the ability for knowledge produced in these tanks to be scaled up to the wider ocean. Debates about the use of reef microcosms continue today even as these tanks became increasingly important to reef conservation. Concerns about the academic rigor of reef tanks might be becoming moot, or muted, by the need for these tanks and the knowledge produced by them to function as spaces to save imperiled wild coral.

It has become apparent in recent years that climate change is changing the ocean more rapidly than the land. Temperature rises, increasing acidification, and rapidly spreading diseases have caused massive coral die-offs and bleaching events that suggest that humans have a limited time in which to lessen or prevent climate disaster in the marine realm. Groups throughout the marine conservation world have sought to bring corals into captivity to both better understand their biological needs and to increase their chances of survival in the rapidly shifting ocean environment.[63]

There are three major roles for reef tanks in conservation efforts. The first is as a space to breed coral that can be planted back into struggling reefs. The knowledge developed by the reef tank community has helped researchers calibrate tank parameters to maintain thriving specimens for reproduction. Coral reproduction is not easy in captivity, and the knowledge gained through the reef tank community has resulted in knowledge not only about coral reproduction in the wild but also about how to manipulate those processes for faster reproduction in captivity.[64] Recently, the Florida Aquarium succeeded with inducing the breeding of several types of endangered coral that are integral to the struggling Florida Reef Tract. The Florida Aquarium, with the help of the Horniman Museum in London and Project Coral, began trying to simulate the natural environment of endangered pillar corals in captivity through the use of water chemistry, lighting, and tidal movement. These corals struggle from an extremely low reproductivity rate in the wild population, making it difficult for the slow-growing coral to recolonize areas where it has become scarce. This problem means that reproduction in the laboratory would greatly increase the chance of this species surviving in the wild.[65] In 2019 they were successful. The pillar coral in the laboratory spawned at the same time as those in the wild. Their success has been repeated for two seasons, meaning that their methods are proving to be replicable.[66] This project hinges on the ability to perfectly reproduce the natural environment in the laboratory. As wild reefs become more imperiled, the breeding of corals in captivity will become an important part of active reef restoration efforts.

The second role that reef tanks play in aiding conservation is by serving as a space for experimentation and observation. As oceans warm, new diseases are spreading around reefs worldwide. One new issue for corals on the Florida Reef Tract is the widespread outbreak of white and black band disease. This disease infects corals and destroys centuries of growth in a very short time. While it is clear that some individuals and whole species are less susceptible to this disease than others, it is difficult in the natural environment to know how the disease spreads. Reef tanks allow researchers to introduce variables—including different animals, various coral species, and varying water flow—to see how these diseases spread. Valerie Paul, the head of the Smithsonian Tropical Research Laboratory in Fort Pierce, Florida, does research on disease spread in her open-air tanks next to the bay in Fort Pierce. Her work will provide a better understanding of how these diseases spread in the wild.[67] Others, such as Madeleine van Oppen in Australia and the late Ruth Gates of the Gates laboratory in Hawaii, hope to use reef tanks to breed and genetically manipulate corals that prove to be resistant to the rising heat and acidity. Experimentation and observation could not be done in tanks without the understanding developed about those systems that was worked out by the aquarist community in the last fifty years.[68]

Finally, reef tanks can serve the conservation community by housing "assurance populations." Assurance populations are organisms taken out of the wild and placed in captivity to assure a genetically diverse population of organisms in the event that wild populations disappear completely.[69] A recent news article detailed a "coral-ark" to save biodiversity in home aquariums using "advanced lighting and water chemistry techniques developed by the aquarium industry."[70] The devastation on coral reefs wrought by climate change means that the assurance populations maintained in home aquariums may eventually be all that is left of their species and what will be the seed of future reefs.

What is clear is that any successful academic or conservation effort with reef tanks will require the continued communication of the entire network of reefers. Bartlett's acknowledgment of public aquarists and hobbyists in his 2013 paper shows how important the craft developed in those groups is to the success of any endeavor with these tanks. If we are going to work on understanding climate change, they must use the most up-to-date craft developed. Conservationists hoping to maintain surviving reefs and rebuild lost ones need the knowledge produced by those individuals who have been tinkering with these tanks for years. In the last decade, we have seen even more links between groups, especially between academic and public aquarists, working together to spread knowledge. Collaborations such as the one between the Horniman Museum and the Florida Aquarium are linking hobbyists, aquarists, and academics into networks that both share craft and push it further to solve the real-world problems of climate change.

As I sat in front of the Adey tank watching the yellow blenny burrowing, I was struck by the idea that this tank might serve as a basis for future reefs. Unlike William Beebe, who was eventually able to differentiate between the simulation of the New York Aquarium and the wild reef, I sat in front of the tank and imagined that this tank was both a past and future reef. Even as we have used tanks to better understand the wider ocean, we have seen those very spaces change rapidly. Tanks have helped researchers understand the basic requirements for building healthy coral, including their photocycles, nutrition, and relationships with other reef organisms. Reef tanks were built to simulate the reef ecosystem as closely as possible, but now they model a reef without peril, one that used to exist and we hope will exist in the future.[71] The knowledge created through these tanks can tell us how to maintain healthy coral in captivity and hopefully build stronger reefs in the rapidly changing ocean. There is power in reef tank simulation that extends beyond developing knowledge about the marine environment; these tanks might be the future of coral reef survival. But corals are only one part of the ocean we hope to understand and save from climate destruction. In the next chapter, we will examine the struggles involved in developing tank craft to simulate a wider range of essential elements to try to breed ornamental fish in captivity.

5

Breeding Tank Systems

Closing the Cycle under Glass

One of my favorite things about doing fieldwork at the St. Lucie County Aquarium in Fort Pierce, Florida, was the time I spent there before it opened to the public each day. On the few days that I showed up before visitors, I was allowed to view the tanks in relative silence as the aquarists performed their early morning routines of tank maintenance. Of course, I visited my blenny friend in the Adey tank and stopped to say hello to the seahorses, but eventually I would find myself sitting in front of a tank that stands out from the rest. The Aquarium is dedicated to educating the inhabitants and visitors of the county on their aquatic resources; the tanks display Atlantic coral reefs, mangrove swamps, and the brackish waters of the nearby Indian River that act as a nursery for juveniles before they make their way into the open sea. Most of the animals in these tanks were caught locally, and in the case of the mangrove exhibit, those young specimens are caught small and released before they get too big. In this way, the aquarium is extremely local in exhibition and impact. But one tank is a bit different.

Placed in an area dedicated to public education and hands-on learning, there is a small aquarium that looks similar to tanks you might see in someone's home. In the aquarium is a vibrant Pacific reef like those built by hobbyists in the last chapter. It is easily contrasted to the murky water of the Adey tank and the dulled colors of the Atlantic mangrove tanks in the rest of the Aquarium.

This tank is the only Pacific reef in the Aquarium, and the purpose of the tank is not to offer an aesthetic juxtaposition to the Atlantic tanks but instead to educate visitors about the importance of captive fish and coral breeding for the health of the oceans. The tank is beautiful, but it is also an incredible feat of tank craft: everything in the tank was bred in captivity. The coral, the fish, and even the crustaceans all came from breeders instead of from wild reefs. The main attraction for me was a small, blue fish with a yellow tail and black markings along its sides: a captive-bred blue tang. This species had only recently been bred in captivity by the University of Florida Tropical Aquaculture Laboratory and was given to the Aquarium to see how it would fare in the tanks. I was looking, each day, at one of the most recent breakthroughs in ornamental aquaculture, one that, it is hoped, will protect tropical reefs from the impact of fish collection for research and display under glass.

This type of tank is rare. Historically, early public aquariums and zoos relied on wild-caught animals for their displays.[1] However, as zoos became more aware of diminishing populations of terrestrial animals, they turned to breeding as a way to contribute to wild populations and to maintain their own exhibits. Zoos and aquariums have been integral to the development of breeding plans for endangered species. The AZA's Species Survival Plans (SSP) share husbandry techniques throughout the conservation network, and institutions involved in these breeding programs share genetic information about their stock to help maintain genetic diversity in captive populations. Some freshwater fishes and marine mollusks, and especially those driven toward extinction through overfishing, have been successfully bred in aquariums. The Tennessee Aquarium works to restock the Tennessee River Valley with lake sturgeon, and a constellation of public aquariums are successfully breeding and releasing white abalone previously fished to near extinction in California waters.[2] But the turn toward this type of breeding endeavor has been slower for ornamental marine fishes, used as pets or exhibits in hobbyist and public aquarium tanks. The use of captive-bred instead of wild-caught fish for display has become an important form of conservation at public aquariums throughout the United States.

The collection of ornamental marine fishes for aquarium display has been on the rise since the mid-twentieth century. Most tropical fish displayed in tanks are caught in the wild, with only about 25 percent of those on display coming from captive breeding. As the number of fish collected has grown, the effect on reefs has become apparent. According to research by the South Pacific Commission on fisheries, ornamental fish collection has affected three separate areas:

> The trade has impacts on three core areas: 1) habitat integrity and biodiversity of tropical marine ecosystems (**sustainability**), including the genetic diversity within individual species; 2) development of coastal communities related to practice safety, economic sustainability, food security, and trade fairness (**equity**); and 3) mortality, morbidity and husbandry of the fishes being cultivated and traded (**welfare**).[3]

The collection methods of ornamental reef fishes, including the use of dynamite and poisons, can be extremely destructive to the marine ecosystem. In addition, estimates suggest that as many as 80 percent of collected fish never make it to market and 90 percent of those sold die within the first year. While most conservation initiatives have focused on tightening regulations and requiring permits and labeling for the market, groups have recently pushed to close reefs to any collecting.[4]

The effects of ornamental collecting on wild ecosystems and the closing of popular reefs to collecting has led to a surge of interest in breeding the most popular aquarium fishes in captivity. However, developing consistent husbandry techniques for ornamental breeds has been difficult. In the 1970s hobbyists and commercial aquarists succeeded in breeding the popular clownfish. Martin Moe, a commercial aquarist and hobbyist, successfully took a generation of clownfish from sexual maturity through breeding and then reared their offspring to maturity and breeding. Successfully rearing a generation completely in captivity and then getting that offspring to breed is referred to as "closing the cycle" in aquarist communities. Shortly after Moe's success, his former coworker, Frank Hoff, demonstrated a scaling up of Moe's work to start large-scale commercial breeding of clownfish. Within a decade, many varieties of clownfish were available on the market, and the aquarium community began working on breeding other species.

But breeding programs stalled for several reasons, including cost and the commercial nature of marine ornamentals. The cost of developing a program for captive breeding was exorbitant compared to the cost of wild fish collection. Some of the first generation of captive-bred ornamentals took over a decade to achieve, meaning that each surviving member might be valued in the thousands of dollars. After Martin Moe closed the cycle on the orchid dottyback, he joked that he *should* sell the three fish produced from the first successful generation for somewhere around two thousand dollars each to recoup his loses.[5] One way to reduce the costs of captive-bred ornamentals is to scale up operations for commercial production. This process sometimes takes years and in some fishes turns out to be too expensive and time consuming. Collecting has conservation impacts, but the lower cost of wild-caught fish and the inability for the buyer to see those negative effects meant that research into breeding stalled. Because most ornamental fish are sourced from the Indo-Pacific Ocean, buyers often cannot see the damage to the reefs left from overharvesting of ornamentals.

However, a network of conservation groups, public universities, state and federal agencies, and the hobbyist community have jump-started a new era of ornamental breeding in the last decade. In the early 2000s, for example, a marine network arose around the conservation of seahorses, a species overharvested for use in traditional medicine. This network linked traditional medical practitioners, public aquarists, academic researchers, and commercial aquaculturists into a group dedicated to working out the craft of breeding these rapidly disappearing animals.[6] At nearly the same time, conservation groups and state and federal agencies began working to breed ornamental fish collected from a range of public aquariums to try to relieve collecting pressure on rapidly diminishing reef stocks. By 2016 over three hundred ornamental species of fish had been bred in captivity, although only a small number have been scaled up for extensive commercial breeding. These successes have seen a renewed interest in closing the cycle on a wide variety of ornamental fishes.[7]

The history of ornamental breeding systems shows the way that tank craft requires deviations between captive and wild populations; in many ways, it reveals the distinctions between oceans and oceans under glass. In previous chapters, we have seen that aquarists start with field observations of wild populations and their environments that they hope to replicate in captivity. Many of the tanks previously discussed work to replicate essential elements from the open ocean as closely as possible. However, the goal of ornamental breeding is to close the cycle of coveted species to provide stock for the wider marine aquarium network. In as much as understanding the native habitats and habits of those species can help with this goal, that knowledge is important to aquarists seeking to replicate the ocean as much as possible. Knowing about the marine ecosystems helps in breeding. But over the course of the history of fish breeding, the requirements of breeding, especially tank design and diet, have been shaped not only by a desire to replicate the native environment but also to achieve a goal with minimal death and maximal profit. In the push to satisfy the market for these fish, the breeding community has also produced knowledge about how previously perceived essential elements can be manipulated to successfully produce thriving fish.

Captive-bred aquarium organisms push us to ask questions about the distinctions between oceans and oceans under glass. Historians and sociologists of science have long been interested in the way that captive-bred organisms are distinct from their wild counterparts. The standardized breeding of animals developed as experimental organisms and systems in science has produced rats, mice, flies, and fish genetically distinct from any in the wild.[8] Agricultural and pet breeders have a long history of choosing phenotypic markers that create organisms with specific behavioral traits and to match desired body size, shape, and aesthetic qualities. Agricultural breeders and fanciers have also produced organisms that are greatly distinguished from their wild types. In the Victorian era, cattle breeders competed to produce the largest bulls, many of ridiculous proportions.[9] Nigel Rothfels's work on the breeding of Prezewalski's horse and its (re)introduction into its native range asks questions about how we distinguish captive-bred animals from those that reproduce in the wild. In his work, Rothfels questions whether the Prezewalski horse bred in captivity for reintroduction in the wild is the same as those it is meant to replicate, and he states that "the modern history of this horse confuses any simple account of what these horses even really are."[10] If a fish is bred in captivity, its environment is a simulation of coral, and it is reared on a diet of man-made food containing the essential elements of its nutritional needs, is it still the same species of fish? In other words, are ocean and ocean-under-glass-bred fish the same fish? While this question might seem like mere academic navel-gazing, these questions are pertinent to understanding how we approach conservation and what we perceive to be the essential elements of the nature we so desperately seek to save.

The development of tank craft around breeding systems also demonstrates how aquarium systems contribute to knowledge about the current and future ocean. Most breeding systems start with knowledge about the food availability and life cycle of an organism; the earliest breeding experiments and systems sought to replicate that environment. But as knowledge about these systems was created, techniques and technologies spread throughout the network. For instance, certain foods have become popular for feeding a variety of species, not because they are available in their natural environment but because multiple species can consume and thrive on the same foods. The popularity of brine shrimp and rotifers stems from their ability to be cultured in large numbers and to be feed to a wide range of organisms. In this way, we can see that tinkering with diet in ornamental systems is both a way to tell what the natural diet of a species is and what deviations their systems can tolerate and still thrive. This process of course mimics the domestication of other organisms. Take the silkworm. Silkworms were traditionally kept in captivity on a diet of mulberry leaves, their natural food source. However, the expansion of sericulture pushed state and commercial researchers to explore alternative food sources and tinker with well-known "regimes of care." The process of tinkering revealed much about the essential requirements of silkworm nutrition and resulted in the development of artificial food.[11] Tinkering with ornamental fish in captivity affords researchers the same insights into the essential needs of marine species. Aquariums are a space that not only tell us much about the open ocean's current inhabitants but also about the potential of those organisms to respond to changing food availability and shifting seas. In this way, aquariums become not just a simulation of what is happening in the ocean but a way for aquarists to model what could happen in future oceans as food availabilities shift.[12]

Finally, studying the development of breeding systems brings into focus several communities that have been largely invisible in the marine network discussed thus far. Both fisheries biologists and commercial aquarists have been largely absent throughout this book. Fisheries biologists perform research in laboratories and the field that directly contributes to the knowledge and health of the state or governing body's marine resources. Commercial aquarists breed fish or create products, such as premade fish food and standardized tank components, that are sold to the larger marine network. Alex Andon's jellyfish tank (chap. 3) is an example of a commercial venture providing tanks, animals, and food. Both communities are important, but unlike other groups, they leave a much lighter paper trail; they are harder to trace than hobbyists. For fisheries biologists, this is because they often work face-to-face with the people who rely on their knowledge; much of the communication of their work is done for the benefit of those constituents. Publication is light and more informal compared to their academic counterparts. Commercial aquarists rarely publish because they hope to use their knowledge for financial gain. This makes it difficult to trace the footprint of commercial aquarists on the larger marine network. Tracing marine ornamental breeding offers a rare opportunity to see the work of these two groups and to examine the variables more closely, commercial and political, that allow or stop the free movement of tank craft.

This chapter explores the history and current practice of marine ornamental aquaculture. The first section details the longer history of both freshwater and marine aquaculture. This section describes the growth of interest from the main groups in this network: state fisheries, commercial breeders, hobbyists, and public aquariums. The second section details the more recent history of marine ornamental breeding with a special focus on two members of the field, Martin Moe and Frank Hoff, as integral figures not only in jump-starting the marine ornamental craze but in spreading their knowledge as widely as possible through speaking, publications, and commercial enterprises. The final section looks at the growth of marine ornamental aquaculture in the 2000s and ends with a sketch of the work currently being done through conservation groups.

The first generation of captive-bred blue tangs were sent to public aquariums throughout the United States. Several were sent to the St. Lucie County Aquarium. On my first afternoon at the aquarium, I sat in front of the tank staring at the blue tangs as they darted throughout the reef. The fish weren't completely blue; when looking at the fish, I could see a lack of coloration in their faces and bellies. The first generation of captive-bred tangs suffered from this distinct discoloration. It was not completely clear to the breeders why that discoloration occurred; it was probably a feeding issue somewhere in the development of the embryo. But the aquarium director was having good luck with returning the pigment through a high-fat diet. As I sat in front of the tank, I became interested in how these fish came to the aquarium and how information about them as they developed further was returned to the breeders. That system of knowledge transfer, and the eventual success or failure of the captive breeding of blue tangs and other marine ornamentals, is what this chapter details.

Aquaculture History

Simply put, aquaculture is the process of rearing aquatic organisms in controlled settings for human use.[13] That definition is more complex than it seems at first glance. There is a wide range of organisms that can be cultured in captivity, including everything from large bluefin tuna to oysters to tiny plankton. The uses for aquacultured organisms are as varied. While we tend to think of aquaculture as the production of food for human consumption, visualizing large salmon pens in the North Sea or trout hatcheries in state facilities, humans culture aquatic organisms for a variety of uses, including fashion (oysters for pearls and mussels for mother of pearl), medical research (zebrafish for cancer research), and of course for the pet trade.

One of the earliest records we have about the purposeful creation of new fish breeds in captivity comes from China and Japan. Records show that Chinese aquaculturists bred carp in captivity to serve for food as early as 3500 BCE. By 500 BCE, carp aquaculture was a robust economy. Fan Lei, a politician turned fish farmer, wrote in his work *Classic of the Fish Farmer* that fishponds were the source of his wealth. The development of carp farming in China eventually led to two aquacultural industries: mass farming for human consumption and ornamental breeding for the pet trade. Even as Chinese aquaculturists domesticated wild carp for food, they began breeding ornamental varieties with vibrant colors and streamlined body shapes to be enjoyed as art by wealthy people in their ponds.[14] Today, these domesticated carp are known as *koi*. During the same period, Chinese breeders also worked to develop a small, ornamental variety with exaggerated features and missing dorsal fins to be viewed not from above, as koi were, but from all angles in glass globes: the goldfish.

Goldfish breeding started in China in the tenth century CE. While the Japanese also began breeding goldfish during this time, genetic studies have shown that all goldfish varieties in China and Japan descend from the wild Chinese carp and not the Japanese, suggesting that the earliest goldfish in Japan were imported from China. Regardless of the origin of the breeding, goldfish became the first ornamental fish bred specifically for its aesthetic beauty.[15] It was not until the eighteenth century that Europeans encountered these fish. In 1780 Edme Billardon de Sauvigny, a French man of letters, published *Histoire naturelle des dorades de la Chine* (Natural history of Chinese goldfish). The manuscript is illustrated with forty-eight hand-colored plates by celebrated natural history illustrator François-Nicolas Martinet. The illustrations are based on a scroll produced by the Society of Jesus, a group that worked in China during the period and passed information about China to Europeans.[16] Goldfish fancying became a popular hobby in China and eventually an important means of scientific research.[17] While it took hundreds of years for goldfish fancying to be noticed by communities outside of Asia, once the craft spread into Europe and the Americas, enthusiasm quickly grew. By the mid-nineteenth century, live fish and craft methods were exchanged throughout Europe and the United States, and this jump-started ornamental fish fancying for aquarium hobbyists.

Both ornamental and food fishes opened economic opportunities for individuals with craft knowledge. In the United States, Livingston Stone and Seth Green started aquaculture businesses concentrating on the production of trout and salmon for market. Green used a combination of written texts and field observations to develop a method of salmon propagation that involved stripping eggs and sperm from mature fish, fertilizing them in jars, and planting the eggs in captivity. Both Green and Stone eventually wrote books on their methods of aquaculture and developed the first network of aquaculturists in the United States. Stone helped to found the American Fish Culturists Association (later named the American Fisheries Society) in 1871, and both he and Green eventually worked for state and federal agencies to try to increase native fish stocks and introduce nonnative high-value food fishes throughout the United States. Stone and Green successfully introduced shad from the Eastern United States into California in 1871.[18]

In the United States the spread of commercial aquaculture eventually led to state-sponsored research in the practice. Green and Stone urged state and federal officials to take seriously the threat of fish depletion to the health of the American people. The federal government developed the Commission on Fish and Fisheries (renamed the US Fish Commission) in 1871. Green worked for the New York Fish and Game Commission, but both he and Stone eventually worked for the US Fish Commission as well. The development of the fish commission

FIGURE 5.1 Illustration of a goldfish variety from the book *Histoire naturelle des dorades de la Chine* (1780). The book is a French translation of an earlier Chinese book on goldfish fancying. Courtesy of the Academy of Natural Sciences of Drexel University.

created an official branch of the government dedicated to the process and science of aquaculture.[19] The first fifty years of the commission involved fractured work; the commission set up laboratories in Woods Hole, Massachusetts; Beaufort, North Carolina; and Davenport, Iowa, to run experiments on the aquaculture of a variety of commercially viable species, including lobsters and oysters at Woods Hole, terrapin turtles at Beaufort, and mussels and trout at Davenport. The scientific studies in these spaces concentrated on understanding the needs of aquacultured organisms, with special attention given to understanding the metabolic rates and food requirements of captive rearing and the identification and alleviation of common pests and diseases in aquacultured species. Many of these experiments involved tank and enclosure development. By 1930, when the United States Bureau of Fisheries[20] declared fisheries biology a new discipline, state, academic, and commercial aquaculturists had developed craft knowledge for culturing a small but consistent list of species, including mussels, salmon, trout, shad, whitefish, and oysters.[21]

Aquaculture also affected laboratory research. In the mid-1930s, a genetics researcher found that Mexican platy fish can inherit skin cancer. Myron Gordon, a Columbia-trained geneticist, received a Guggenheim fellowship to set up a breeding center for these fish at the New York Aquarium (NYA). This breeding program supplied fish with skin cancer to laboratories throughout the United States. When the NYA closed for renovations, the tanks shifted first to the American Museum of Natural History and then in the late 1990s to Texas State University, where it is now known as the *Xiphophorus* Genetic Stock Center.[22] The zebrafish became popular with cancer researchers in the 1990s. While many researchers would like to use fish in the laboratory because they represent another vertebrate on which they can experiment, the difficulty of finding space for all the tanks required and the craft required to maintain them means that very few fish are cultured in laboratories as experimental models. Both platyfish and zebrafish are kept in fairly simple, automated systems. But even those automated systems require tank craft from veterinarians, researchers, and lab technicians to maintain large numbers for the long term.[23]

The early focus on freshwater species was not only because these were considered economically important but also because they proved easier to culture than their marine counterparts. Marine species proved particularly difficult to cultivate. Many freshwater species spawn in moving water, but the juvenile forms can live and mature in largely stagnant pools or ponds. Seth Green's methods for hand fertilization, developed in the mid-nineteenth century, meant that eggs could be matured in small hatching jars about the size of a two-liter bottle and then reared to juveniles in small tubs or ponds before being released into larger pools to

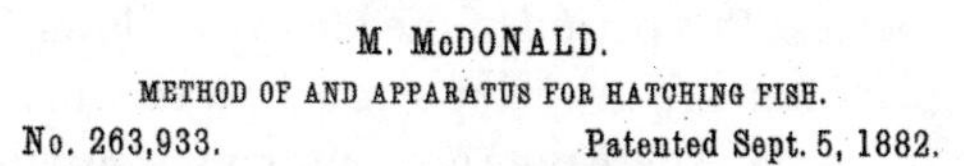

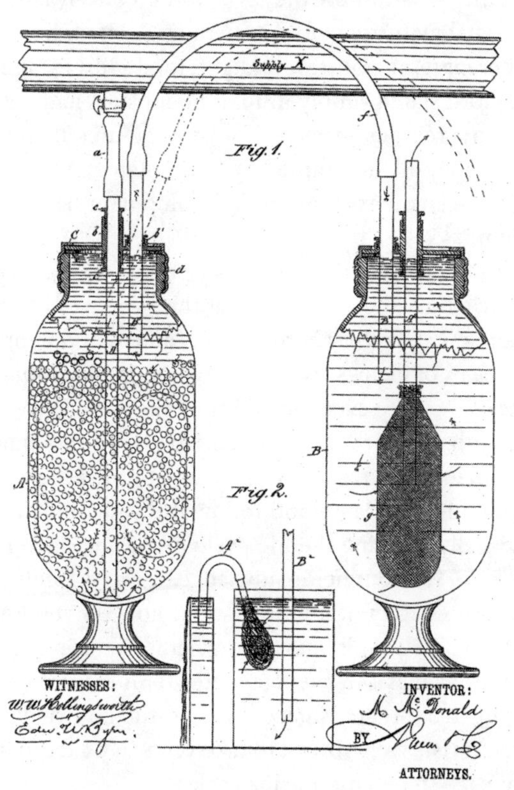

FIGURE 5.2 A popular hatching jar known as the McDonald jar. Fertilized eggs would be placed into these jars to mature. The US patent was granted in 1882. #US263933A, https://patents.google.com/patent/US263933A/en.

achieve maturity. A freshwater hatchery could consist of a room lined with hatching jars hooked to a water source and a series of ponds containing fry at varying sizes. This meant that these juvenile fish spent very little time under glass. Marine species proved unable to be successfully controlled in captivity in these ways. The earliest interest in lobsters showed how difficult rearing in captivity would be.

Initially, aquaculturists tried to collect sexually mature female lobsters; the hope was that these females would then immediately spawn in captivity. But spawning in captivity was only the first step in rearing lobsters to maturity. Young lobsters are tiny organisms called gelatinous zooplankton—once hatched they swim in the water column and mix with other juvenile species. This period of the life cycle allows the lobsters to prey on small plankton and grow big enough to eventually settle onto the bottom of the ocean, where they grow progressively larger and eventually reach sexual maturity before spawning. Each of these areas of the ocean has different temperatures, sunlight, food sources, and structures. As D. R. Crawford stated simply in 1921, no lobsters had been reared in captivity, and "the reasons advanced for these failures are numerous and varied, but they may all be summed up in the statement that the natural conditions for the larval existence and development have not been met in the aquarium."[24]

Replicating the ocean in the laboratory proved tricky. One major difficulty of culturing marine species is that the variety of life cycles made following them in nature very hard. The ocean environment is vast, and the different stages of many species live in differing locations during the various stages of their life cycle. In addition, these life cycles often don't resemble each other, and they are difficult to follow. When near the surface of the ocean in their juvenile stage, even a trained eye would be hard pressed to spot a lobster larva and follow it through its life cycle. This means that visually tracking the life cycle of marine organisms proved nearly impossible. Without field knowledge, it was difficult for aquaculturists to know how to craft a tank system: how do you know how to house something if you can't figure out what it needs in the wild? What does it eat, and what is the average water temperature of its juvenile range? Without this direct visualization and the knowledge that came with it, aquaculturists were forced to take years to tinker to find optimal conditions for each step of marine life cycles.

For marine organisms, aquaculturists had to try to mirror the natural environment as closely as possible. But because that environment was largely inaccessible, it took a feedback loop of tinkering with captive conditions and then checking them against natural conditions until all steps in the life cycle could be worked out. In this way, we clearly see how building tank systems creates knowledge to shed light on marine ecosystems and simultaneously pushes tank craft forward.

Ornamental Aquaculture

Aquaculture was not only the province of commercial and state enterprises. By the early twentieth century aquarium hobbyists began experimenting with the breeding of a wide range of freshwater ornamentals. Of course, the goldfish was usually the first and the most popular fish to be bred by hobbyists.[25] The exportation of goldfish varieties from China and Japan jump-started fish fancying worldwide, with hobbyists breeding goldfish from Berlin to San Francisco. These groups built on the earliest variations worked out, including differing coloration, eye shape, fin shape, and body size. A black telescope, denoting a black goldfish with protruding eyes (like telescopes), might be shown or traded at a hobbyist club for a Ryukin, a short bodied, hump-shouldered fat fish, or a calico fan tail, a multihued fish with a long, elegant tail. Awards were given for the best fishes. Between 1908 and 1917, William Innes garnered awards at the Aquarium Society of Philadelphia for his scaleless fringetail, scaleless nymph, and scaled telescope. These varieties were usually produced by serious hobbyists and traded among themselves. The commercial goldfish industry produced the simpler goldfish which still serve as popular first pets for children worldwide. In 1917 Innes published a hobbyist manual on goldfish fancying and keeping that introduced many of the fancier goldfish to a wider audience.[26]

Breeding marine tropical ornamentals for the aquarium industry lagged far behind freshwater species. It was not until the late 1960s that a breakthrough with clownfish signaled the development of a new phase in ornamental aquaculture.

Clownfish

The development of clownfish aquaculture clearly demonstrates the inextricable connections between commercial, state, public aquarium, and hobbyist aquaculture. In 1966 Frank Hoff, a recent graduate from the University of South Florida (USF), was hired to work with Martin Moe at the Florida Department of Natural Resources Marine Research Laboratory. Martin Moe had earned his undergraduate degree from Florida State University and a master's from USF before beginning his work in the department. Hoff and Moe were initially assigned to work on taxonomy and migration of the fish of the Gulf of Mexico. According to Hoff, their boredom with working as "virtual morticians" surveying dead specimens pushed them to start a side project with black seabass. Both Moe and Hoff had marine biology backgrounds, but this was their first attempt at breeding a marine species. Hoff describes these results as mixed: they never achieved fertilization of eggs without stripping the sea bass and hand mixing the eggs and sperm. They built a large larval rearing system that replicated the motion of the water column, but they didn't succeed in rearing seabass in the tanks, although they did successfully grow out several species caught in the water column during plankton tows.[27]

By 1970 Martin Moe left the Florida Department of Natural Resources to start a commercial breeding program spawning and rearing pompano at his company Aqualife Research. Pompano is a large marine fish, popular with anglers. Moe was not the first to become interested in pompano. In 1967 R. H. Lewis began working on pompano research for the Minorcan Seafood Company of St. Augustine, Florida. According to Moe, "this research was harried by limited funds, time, and equipment, and was prematurely terminated."[28] But Lewis shared his data sets with Moe and Robert Ingle in a publication on pompano mariculture for the State of Florida in 1968. In the acknowledgments of this publication, we can see the intricate weave of state and commercial interests for aquaculture. The authors thank the Bureau of Commercial Fisheries, Miami; the Institution of Marine Science, University of Miami; Florida Seafood Growers, Pensacola; Minorcan Seafood Company, St. Augustine; Auburn University, Auburn, Alabama; the Bureau of Commercial Fisheries, St. Augustine; Groton Associates (later Groton Bioindustries, an aquaculture technology think tank); and several individuals.[29]

After working out the parameters and needs of the pompano, Moe turned his attention to clownfish. By the 1970s clownfish were the most popular tropical marine fish for hobbyists worldwide. The clownfish is an orange, black, and white marine reef fish originating from the Indo-Pacific region of the world. Besides their bright coloring, the clownfish was also popular with those keeping reef tanks with anemones because they live symbiotically with anemones in the wild.[30] However, most clownfish were collected using cyanide. To collect fish this way, cyanide tablets are dissolved in salt water and then sprayed on the reef as close to the fish being collected as possible. The cyanide stuns the fish long enough for it to be collected by hand. While cyanide makes collection easier, it is indiscriminate and often hurts the fish, meaning that many clownfish arrived at pet stores already weakened and distressed. They are also susceptible to a host of common parasites. There was a high mortality of wild-caught specimens. The demand coupled with these well-known issues with commercial wild-caught supplies gave Moe the idea to work with clownfish to develop a tank-bred market. In 1973 he published his first major hobbyist article in *Salt Water Aquarium* on successful aquaculture of the clownfish *Amphiprion ocellaris.* The all-caps headline of the magazine proclaimed "TANK RAISED CLOWNFISH! SPECIAL SPAWNING ISSUE. A MILESTONE IN THE HOBBY."

In the article, Moe described his process of setting up spawning and rearing tanks. Moe set up two fifty-five-gallon spawning tanks meant to hold mating pairs. From the beginning, he tested variables in each system: he populated one tank with natural seawater and used small local fish to provide nitrogenous waste to activate the filter bed. In the other he used artificial seawater and activated the filter bed with ammonium chloride. Over time, he found that the second tank was much easier to control; the natural seawater and local fish introduced parasites into the water that required several antibacterial treatments over the course of the first experimental phase. Each tank contained anemones and concrete habitat for egg-laying substrate. A single mating pair was introduced in each (although one pair turned out to be two females).

Breeding and rearing aquatic organisms requires at least three tanks: one for mature adults to spawn, one for newly hatched larva, and one for growing juveniles into maturity. The spawning tank contains sexually mature fish and simulates the type of environment—including temperature, photoperiod, and habitat—that a mated pair would find in their natural environment. Clownfish use anemones to protect their eggs in the wild, usually anchoring them to the coral substrate on which the anemone is living. The rearing tank must be separate from the spawning tank for several reasons. One is that it isn't uncommon for parents to feed on their own offspring; removing recently hatched larva prevents this predation. In addition, newly hatched fish require small food particles and open water; most tropical fish subsist in the pelagic zone with other gelatinous zooplankton. This gives them an open area where they can prey on a variety of microscopic food while the current in this zone keeps the larvae floating and able to capture food without expending too much energy. The larval tank simulates this environment using similar methods to the kreisel tank in chapter 3. As they mature, juveniles move lower in the water column and develop secondary sexual characteristics such as coloration. At this point they can be transferred to a grow-out tank that contains fewer individuals and allows the fish room to grow and develop. It is in this space that clownfish often find mates (they mate for life). All clownfish are born male (protandrous hermaphrodites), meaning that they shift from male to female. The rearing tanks must closely model the natural environments in which these behaviors occur, and the aquarist must pay attention to exact moments of spawning, hatching, and when larvae transition to juveniles so that they can provide the proper tank environment for that portion of the life cycle.

Moe's system of breeding clownfish relied heavily on close observation and tinkering. Initially, he thought to take the eggs out of the anemones right before they were supposed to hatch (reported as eight days after spawning in the wild) but this resulted in poor hatch quality. Instead, he left the egg ball in the anemones and siphoned newly hatched larvae out of the spawning tank immediately upon hatching. After fifteen to eighteen days in the larval tank, floating and being fed wild-caught plankton and successively bigger foods, the juveniles that survived were placed in the grow-out tank. This sixty-gallon tank contained anemones, rocks, and a sandy bottom. The fish developed coloring and sexual characteristics in this tank. Moe closed the short article by claiming that while his successes were small, his system had possibilities for the production of a wide variety of marine tropicals.[31]

While Moe was working on clownfish, Frank Hoff was also transitioning into this research. In 1974 Frank Hoff moved from state aquaculture to private industry with a focus on clownfish production. Instant Ocean Hatcheries (IOH; initially named Neptune's Nurseries) was first located only three blocks from Moe's facility and only seven blocks from the Marine Research Laboratory (MRL) where both started working with pompano years earlier. The funding for the endeavor was provided by a combination of investors with ties to the Mystic Aquarium and Cleveland Aquarium as well as commercial aquarium enterprises. Hoff hired one of his colleagues from the MRL and his brother to help him start a large clownfish breeding program at IOH. Moe and Hoff were friendly rivals, both working to set up competitive breeding facilities while sharing information; the two companies were headed by two men who worked together and were the leaders in tropical culturing.[32]

Even though Moe had already managed to rear several generations of clownfish into sexual maturity, there was space for Hoff to expand and perfect the breeding systems. By 1976 IOH had succeeded in culturing six species of clownfish and had reared the first known tank-raised marine hybrids: *A. frenatus* crossed with *A. ephippium* and *A. frenatus* crossed with *A. akallopisos*. By 1977 IOH added three more species, a neon goby, and another hybrid. Neither Moe's nor Hoff's successes with culturing equated to commercial success; both companies eventually went out of business. However, when the commercial enterprises failed, Moe and Hoff began publishing more widely about their techniques, resulting in the dissemination of the previously closely guarded commercial secrets.

By the late 1990s Hoff and Moe began writing manuals on tropical fish culturing for both commercial and hobbyist audiences. Hoff's 1996 *Conditioning, Spawning and Rearing of Fish with Emphasis on Marine Clownfish* was "dedicated to the avid aquarist and beginning aquaculturist. . . . My desire is for the reader to understand as much as possible. Therefore, I have purposely avoided extensive scientific verbiage yet provide professional rather than popular amateur discussions."[33] According to Hoff,

> Even though data provided in this manual was recorded over 30 years ago, it still stands as the only complete commercial data available to the public. To date, only limited, ambiguous technology transfer has been provided via hobby literature. I hope this manual provides your door to the future and will eliminate the need for future investors to reinvent the wheel. Why waste a great deal of time or effort in creating something

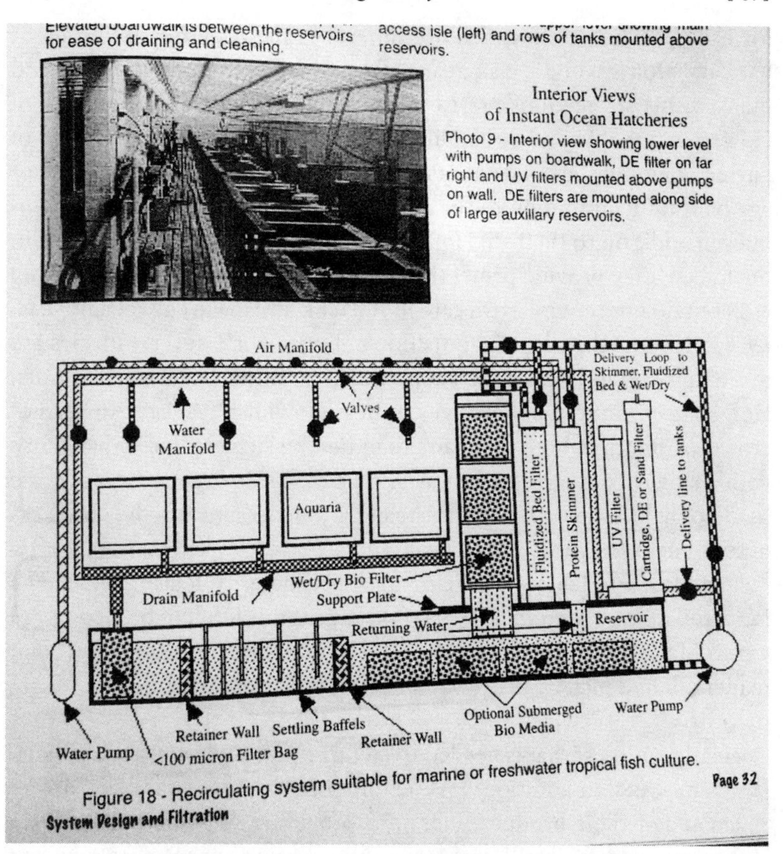
Elevated boardwalk is between the reservoirs for ease of draining and cleaning.

access isle (left) and rows of tanks mounted above reservoirs.

Interior Views
of Instant Ocean Hatcheries

Photo 9 - Interior view showing lower level with pumps on boardwalk, DE filter on far right and UV filters mounted above pumps on wall. DE filters are mounted along side of large auxillary reservoirs.

Figure 18 - Recirculating system suitable for marine or freshwater tropical fish culture.

Page 32

System Design and Filtration

FIGURE 5.3 Page 32 from *Conditioning, Spawning and Rearing of Fish with Emphasis on Marine Clownfish* (1996). Unlike Moe, Hoff's work and images were meant to help the reader set up a commercial clownfish operation. His images are of many large tanks and scaled-up filtration systems that will work for multiple varieties of fish. Courtesy of Florida Aqua Farms.

that already exists. New investments into marine tropical fish should only add to our current knowledge pool and not be wasted.[34]

Hoff makes it clear that the book would not, or perhaps could not, be written if IOH was still in operation. Because it was not, he felt it was important to convey "to future farmers technology and data which has been well paid for and does not have to be repeated."

Hoff's book shows how far clownfish breeding had come in twenty years. Moe's original article was a mere six pages, and Hoff's book, including citations and water quality data, is over two hundred. While much of the system described by Moe had not changed, the parameters for each species of clownfish are different, and they require different habitats. Hoff's work is also geared toward large-scale production, and many of his suggestions are to make production of large numbers of clownfish possible and streamline the process. For instance, instead of having anemones in spawning tanks, which makes retrieval of larvae or egg balls difficult to monitor or remove and requires care for the anemones in addition to the fish, Hoff suggests placing red clay flowerpots in the tanks. They provide protection for the fish and a substrate for laying eggs but do not require extra care in the tank and make egg retrieval easier. Hoff warns that the clay pots don't always work: several species are more nervous and prefer the anemone. In addition to the red clay pots, Hoff suggests more tank changes to help clownfish feel safe. Instead of crowding multiple breeding pairs together, he suggests individual tanks with a single front facing window and opaque sides painted a moderate shade of light blue with soft overhead lighting to simulate the closeness of rock enclosures and reef environments.[35]

In addition to his book on clownfish breeding, Hoff also published a *Plankton Culture Manual* in 1987. Hoff meant this book to be a companion to his work on clownfish. In his work on clownfish, he stated that readers should pick up his *Plankton Culture Manual*, which

> clearly outlines what is needed to culture microalgae, rotifers, daphnia, artemia, etc. In addition, FAF (Florida Aqua Farms, Inc) has further expanded their product lines into providing "off-the-shelf" starter cultures, special fertilizers, and culture equipment to make your job easier. . . . In order to provide a balanced diet, and to develop adequate transition from live rotifers to live *Artemia*, a series of dry micro particle foods must be used throughout the rearing process. Rotifers, *Artemia* cysts, microalgae, and micro foods are all available from Florida Aqua Farms Inc.[36]

One result of increased aquaculture of marine ornamentals was that aquaculturists needed access to a wide range of food relatively quickly. Fish larvae floating in the water column have access to a wide range of prey throughout their growth cycles. Many newly hatched marine ornamentals have extremely small mouths and do not have the ability to eat large prey. However, they grow rapidly, and within a few days to a week they need a completely different diet to continue normal growth. Figuring out the feeding schedule for new breeding is incredibly difficult, especially because Moe has observed that clownfish have differential growth rates. Having larvae from the same brood requiring multiple food sizes can be difficult. To solve these issues, most tropical aquaculturists began culturing multiple food items in-house.

Hoff's book, cowritten with Dr. Terry Snell, at the time a marine ecology professor at University of South Florida, provided both ecological and aquacultural information about food species. The manual was meant to "provide the basis and principles of live food culture." Over the course of thirteen chapters, Hoff and Snell walk hobbyists and potential commercial aquarists through the role of food sources (daphnia, plankton, clams and oysters, and copepods) in the ocean food web, explain their reproduction cycles, and detail the systems required to culture these food items in captivity. Hoff states that to have enough food for breeding, you need at least a rotifer and plankton rearing system in addition to purchased larger food items. His book shows the expansion of the breeding system. Moe's initial system sounds somewhat simple: three tanks with natural elements. However, by the late 1980s and early 1990s, ornamental aquaculture required at least three tanks (which could only accommodate one or two hatches depending on growth rates) and a wide range of tanks to grow enough food to keep those different life cycles of the alive and thriving.[37]

Both Hoff and Moe were aware of the way that knowledge used in commercial enterprises might be cut off from the larger tank craft network, and they worked conscientiously to break down this wall. In 1997 Moe published *Breeding the Orchid Dottyback: An Aquarist's Journal*. This book is a lightly edited laboratory notebook kept by Moe from February 1996 to March 1997. It details his attempts to breed the orchid dottyback in his home laboratory (a converted half bath). Moe was in direct contact with both Hoff and Julian Sprung during this period of his work, and he directly references their help in building this breeding program. Over the course of a year, Moe worked out the breeding and feeding cycles of the orchid dottyback. The beauty of this book is that Moe was incredibly honest about the time, space, and monetary outlay required to figure out how to breed these animals.

Moe's work shows that commercial food and products did not cut out the tinkering required for new species breeding. Seeing that the larvae struggled after the first week, Moe found that the trouble seemed to be that larvae could not eat enough of the rotifers he was initially feeding them: only a few of the rotifers were small enough to be eaten by the larvae, and the others quickly took over the tank. After visiting with Julian Sprung (chap. 4), Moe decided to start collecting wild plankton cultures for food. Using Sprung's plankton net, Moe began collecting plankton several times a week (sometimes more or less depending on the season and quality of the tow) and then sieving the smallest plankton into the tank.[38] This resulted in growth, and Moe was able to eventually rear larvae into the juvenile and mature stages. Eventually, the feeding schedule for the fish included plankton for the smallest larval stage, and then grated and sieved frozen shrimp bits and rotifers, followed by artemisia and larger shrimp. The rotifers are particularly interesting in this case. Moe found it very difficult to keep his rotifers alive because his algae (their food source) would not grow nicely in his home tanks. Without the algae for food, his rotifer cultures were sparse. While eating breakfast one morning, Moe had the idea to introduce V8, a tomato-based beverage, as an alternative food source for the rotifers, and this worked well enough that he included instructions for his V8 rotifers at the end of the book.[39]

Both Hoff and Moe developed breeding systems by balancing knowledge about the natural history of their subjects with the constraints of a system that only simulates the essential elements of the environment. Hoff made it clear that a new breeding endeavor must be preceded by research. "To determine the parameters for such a regime, it is imperative to seek as much data as possible about the fishes' normal environment and habitats. Weather bureau and coast guard literature will provide a wealth of routine data on seasonal water temperatures. Daily light periods are available for various regions from the weather bureau or various governmental agencies."[40] Moe noted in his update to his 1973 paper that "The thing that always helped me with development of culture procedures with new species was to learn as much as I could about the natural history of the species under culture, the environment, the diet, the reproductive modes, etc., and use that knowledge as a base for development of adequate substitutes for that organism's basic requirements for survival."[41] Both Moe and Hoff had a background in academic biology (they both hold master's degrees) and worked in fisheries research before starting their endeavors with commercial breeding. Their books contain a large number of citations from academic journals on the natural history of the fish they were working to breed. However, this did not mean that they worked to provide completely natural conditions for their brood stock.

Because no tank system can perfectly replicate the natural environment, breeders seek to find what natural variables are indispensable and which others can be modified to suit the desired outcome. Moe started his clownfish project by including what seems to be one of the most important aspects of their habitat: the anemone. Moe used anemones in his tank design and included them in his article on breeding, but Hoff eliminated this aspect of the tank. By introducing the clay flowerpot, most breeds of clownfish received the positive aspects of the anemone, including protection and egg-laying substrate, without the breeder also having to maintain the anemone, which require light, food, and current conditions that are similar but not necessarily as variable as for the clownfish themselves. Hoff also stated that clownfish mate for life, but couples are often *trios* in the wild. Commonly, a male and female welcome a smaller male into their company. This is probably because females are usually older than their mates and expend more energy on reproduction, and consequently they die first. When the female dies, the larger male transitions into a female, and the smaller male becomes the male of the pair. This assures the clownfish that there is always a mate for reproduction. Hoff explains, "In nature, this is a unique adaptation for continued survival of a pair. Yet, in commercial operations, I do not feel it is good practice to leave an extra fish in the tank since it interferes with normal hatchery practices." Hoff instead suggested that when the female dies in the hatchery, a young male should be introduced to the surviving male. While it takes up to several months for the older male to transition to female, it still occurred, resulting in viable offspring from the surviving member of the original breeding pair.[42]

The craft work in breeding comes through the negotiation of natural and artificial elements to produce the desired outcome: captive-bred ornamental fish. During the clownfish breeding experiments, both Moe and Hoff started with information about the natural habitats and behaviors of that species. But Moe seemed to work from the other direction when beginning his orchid dottyback experiments. He initially fed them lab-cultured or store-purchased foods but ended up with a system that required wild-caught plankton and V8-cultured rotifers. Additionally, Moe negotiated the importance of allowing the male dottyback to maintain the egg mass. In the wild, the female dottyback deposits the egg mass and then leaves it for the male to fertilize and protect until the eggs hatch. Initially, Moe's male dottyback ate the brood, either right before or during hatch. Moe tried to move the mass into another tank for hatch but found that they did not develop correctly. Instead, he found over time that for successful brood and hatching, the egg mass should remain with the male even if this meant that some stock was lost to parental predation. Moe suggests this might be because the male "mouths" or "fans" the eggs during development and hatching. He claims that he did not see distinguishable behaviors during hatching but that the result is obvious. Instead of trying to get around this behavior by creating an artificial environment, he accepted this as part of the natural behavior of the fish that must remain in the aquaculture process.[43] Both Hoff and Moe worked from details of natural history but also observed and recorded new behaviors of their subjects. This feedback loop—using the model system to tell you how important the variables of the natural system are to achieving the outcome desired—means that both Moe and Hoff contributed knowledge not just to the aquarium community but also to understanding of these fish in the wild.

The lack of commercial viability in aquaculture of marine ornamentals—the difficulties in inducing reproduction, maintaining large numbers of larvae, and the significant cost of having to build such an extensive system of food and habitat for individual species—hobbled the industry. Both Hoff and Moe spent years perfecting their craft but gave much of their initial knowledge away after commercial endeavors faltered. While working on the orchid dottyback, Moe joked that

> This marine fish rearing is a very time-consuming business. Actually, it's not a business because I could never get $5,000.00 each for the fish. I don't think anyone with a full time job could ever rear these things, even just experimentally. Clownfish, yes, they are much less demanding of time and attention. But then, on the other hand, once a routine system for rearing the dottybacks is developed, when one knows exactly what is required and when, then it may not be quite as consuming of time and effort to rear them.[44]

Instead, he brought his dottybacks to hobbyist club meetings and offered them for sale at a very low price and as prizes. But the true gift that Moe and Hoff gave to the community was their open communication of the craft knowledge they developed. It continues to be important to the industry today.

Bringing More Breeding under Glass

Ornamental reef fish have seen a sharp decrease in wild stocks over the last half century, and collecting for hobbyists has abetted this decline. One of the first communities to work on breeding in order to both supply markets and preserve species viability were those working with the seahorse. Project Seahorse was developed by American biologists Heather Hall (now Koldewey) and Amanda Vincent and environmental anthropologist Marie-Annick Moreau in 1998. Seahorses belong to the family *Syngnathidae* of marine finfish that include pipefishes and sea dragons. Traditional Chinese and Indonesian medical practitioners have used seahorses in their medical practices for centuries. Added to this, the hobbyist and aquarist community has fallen in love with these creatures. In many public aquariums, seahorses attract attention, and they continuously rank among the most popular species. Much of the interest in seahorses comes from the anthropomorphization of their pair-bonding and the fact that the male carries and births the live young. Their use in medicinal practice and aquariums has placed pressure on dwindling populations, struggling to survive amid habitat encroachment and warming seas. By the mid-1990s, those specializing in seahorse populations saw a dramatic drop in wild populations because of the stresses of commercial and hobbyist pressures. While exact numbers were difficult to come by at the time, estimates of decline by those studying them in the field were dire enough to induce the International Union for the Conservation of Nature to list all species on CITES (Convention on the International Trade in Endangered Species) appendix 2 as "endangered" because of habitat encroachment and unsustainable collecting practices. Between 2001 and 2004, monitored populations dropped between 79 and 98 percent.[45]

Project Seahorse brought together a wide network of interested parties to develop seahorse aquaculture. In July 1998 a diverse group including traditional medicinal practitioners, aquarists, and academics met in Cebu City, Indonesia, to discuss the possibility of breeding seahorses. Hall, Vincent, and Moreau stated that

> Many fishers, traders, and traditional medicine practitioners had been asking for help to balance dwindling supplies with growing demand for marine species of medicinal value. Nearly all these producers and consumers suggested that aquaculture could help meet needs. The aquaculturists, meanwhile, were acutely conscious of the difficulties in culturing many marine species. They also wanted advice on which individuals and which species were most marketable. Fisheries managers and conservationists were becoming increasingly concerned about overconsumption of marine species for traditional medicine, and wary of aquaculture ventures that might do more environmental damage than good. It seemed time to talk, and find common ground for future initiatives in managing marine medicinal species.[46]

This meeting was novel for many reasons, but one important one is that it included direct communication between those who might be able to culture seahorses with both groups of individuals hoping to exploit them for medicine or aesthetics. One goal of the meeting was to have medical practitioners develop a list of their most-sought species so that aquaculturists might know which species should be cultured first. Hall and Vincent built a community where they acknowledged the use of seahorses in traditional medicine and then sought to help with the conservation of these organisms, not by telling practitioners not to use them but by bringing the problem to aquariums and aquaculturists and asking for help. Medical practitioners not only told the community what they most desired but also where they found them, when they harvested, and any information they had about the natural habitat of many species that hadn't been fully studied by aquarists and academics.

The inability to understand the behaviors and growth of wild seahorses inhibited the initial breeding programs. Seahorses are generally difficult to track for extended periods in their natural environments. Many species live in unclear water or hide in large seaweed patches for safety. For this reason, it was not until the latter half of the twentieth century that anyone even knew that wild populations might be declining. This lack of visibility also made it difficult to develop a framework of needs for captive care and breeding. In particular, understanding the feeding requirements of individual species proved difficult. Newly hatched seahorses require an abundance and variety of food that shifts rapidly over the first weeks of their life. While they are more capable of taking large foodstuffs than Moe's orchid dottybacks, their nutritional requirements mean that volume and consistency of feedings place a burden on the aquarist. While some aquarists have sought to provide wild-caught food, because of these constraints, most seahorse culture is done with purchased or cultured amphipods, decapods, and mysids. Seahorses prefer live prey, so while some can be convinced to take frozen shrimp or other frozen and thawed foodstuffs, it is easier to provide living food even though this takes extensive space for culturing.[47]

In addition, seahorses have intricate bonding and mating behaviors, resulting in lifelong monogamy. Replicating these activities in captivity proved difficult at first. Unlike the clownfish, who could be sexed and placed in mating pairs after maturity, seahorses have a more protracted process of forming mating pairs, with males choosing females of larger sizes and females choosing males based on pouch size and general dancing characteristics. Allowing for these mating behaviors means building space and time for them to occur in the tank system. As seahorses move from their open-water to benthic stage, they wrap their tails around plants. Deciding when to introduce structures to the tank and what materials will make maintenance of the tank easiest is another hurdle for breeding. Finally, seahorses suffer from several illnesses in captivity, including bacterial infections and sensitivity to gas bubbles becoming trapped in the brooding pouch or subcutaneous skin layers. It is easier to try to prevent these issues by treating food with antibiotics and giving care to oxygen levels, because their bony plating makes it difficult to administer needle-based antibiotics.[48]

In many ways, Project Seahorse has been an aquaculture success.

When the initial meeting occurred, there were few aquaculturists breeding seahorses in captivity, and those operations were largely hobbyists or for small scientific research units. Since 1998 public aquariums and commercial endeavors have developed aquaculture methods and by 2005 thirteen species of seahorses were being commercially cultured for sale on the aquarium market with another two cultured for research. Public aquariums are especially engaged with this aquacultural enterprise, with a network of Association of Zoos and Aquariums (AZA)-approved institutions breeding and trading seahorse species. Project Seahorse was originally centered at the Shedd Aquarium in Chicago, Illinois. This institution continues to breed seahorses, but the endeavor has expanded to almost every major aquarium. The building of the Project Seahorse network made it possible for fishers and traditional medical practitioners to share their knowledge of wild populations with aquaculturists and aquarists. While large-scale breeding operations are still in their infancy—and most cultured seahorses are sold to hobbyists or aquariums, not into the medicine market—it is hoped that this endeavor will protect wild populations in the future.[49]

The successes of breeding clownfish and seahorses, however, did not translate into success with other species. By 2016 breeders succeeded in successfully culturing about three hundred species of fin fish in captivity. However, less than thirty have been scaled to commercial production. This lack of commercial production is largely due to the inability to standardize breeding operations because of the spatial and temporal requirements specific to marine species. Each species of fish requires at least three grow-out tanks and at least three different live cultures for feeding. To produce a single generation of captive-bred fish, that means at least six tank systems and the corresponding focus on water quality, filtration, and time input for cleaning, sterilizing, and other forms of maintenance. But to maintain multiple hatches, you need an enormous amount of space for housing and feeding fishes in different stages of maturity. As Moe noted with the dottybacks, slower development was not an indication of illness or lack of viability, meaning that it behooves the breeder to maintain the slower developers because they will be viable. But large-scale productions rely on consistent development and the ability to manage those populations accordingly. Marine ornamentals have proved very difficult to scale up in production, and therefore progress has been slow.[50]

Yellow and Blue Tangs

As wild populations dwindle, those dependent on the ornamental trade have turned to the development of aquaculture for the most popular species. Hawaii collects and exports the most ornamental marine fishes for the American hobbyist trade, and this group of fish became the epicenter of ornamental captive breeding experiments at the turn of the twenty-first century. The Oceanic Institute (OI) is an affiliate of the Hawaii Pacific University (HPU). The institute is an experimental aquaculture laboratory dedicated to the development and dissemination of economically viable aquaculture technologies and techniques throughout Hawaii. In the early 2000s the OI began focusing on the captive breeding of three of Hawaii's largest ornamental exports: flame angelfish (*Centropyge loriculus*), Potter's angelfish (*Centropyge potteri*), and yellow surgeonfish (*Acanthurus xanthopterus*). Over the course of almost twenty years, the lab has successfully bred all three.[51]

The OI started working to breed these fish in the late 1990s, and it took over a decade to close the cycle on these three fish. The difficulties with the flame angelfish demonstrate the wide variety of tinkering required to breed in captivity; researchers had to pay attention to tank structure, population density, and of course, food sources. The flame angelfish is beloved by hobbyists for its intense coloring, and it is one of the most popular aquarium fish. While it is common for these fish to spawn nearly every day, and for hobbyists to view the spawning in home aquariums, it proved difficult to get fertilized eggs to develop. Researchers found that not all eggs are created equally; to get quality offspring, the spawning pairs must have optimal conditions, including the right food and space. The OI found that while they could obtain spawn from heavily stocked tanks, they received more viable and hardy eggs by placing a single mating pair in large tanks, devoid of the more common harem seen in the wild. In addition, the newly hatched larvae proved to have extremely small mouths, necessitating a search for a suitable first food. After much experimentation, a single species of copepod, *Parvocalanus* sp., was found to be of an acceptable size and nutritionally sound enough for that first food.[52]

While the OI was working out the cycle of these fishes, an independent aquaculturist in Hawaii, Frank Baensch, successfully closed the cycle of three pygmy angelfish species. Baensch moved to Hawaii after he studied aquaculture in graduate school and used his cultured angelfish to fund the development of a private company, Reef Culture Technologies. Between 2001 and 2011, Baensch closed the cycle on many angelfish. His initial first foods came from wild-caught plankton. But his methods beyond this are difficult to track down. While Baensch has a beautiful website, he has only written a few articles about his breeding process, probably because his commercial success depends on his ability to produce these species before or more consistently than his competitors.[53]

After the breakthroughs with angelfish, the potential of a wide variety of popular aquarium fish for culture became evident. The group Rising Tide, a nonprofit organization funded by SeaWorld, began funding laboratories in Florida and Hawaii to work on closing the cycle on popular aquarium fishes. The goal of the organization was to create networks to develop and share information about the breeding of these specimens. Work on the yellow tang began at the OI in the 1990s, but success was hampered by low egg quality and an inability to find suitable first foods. Finally, the yellow tang was successfully cultured in captivity in 2015 using overlapping feeding of three separate foods over a seventy-day period. This success was trumpeted in the popular press, most likely because of the popularity of the species. The yellow tang is one of the most popular aquarium fish in the hobbyist community and the largest export of the ornamental industry in Hawaii. But it also came at a time when conversation and political debates about closing that fishery began to solidify. The ability to breed yellow tang was previously a problem to be solved for conservation or possible economic purposes, but by the time the first generation of juveniles matured, they became a symbol of the birth of an industry that might help Hawaiian fisherman recover from the possible closure of their reefs to ornamental fish harvest.[54]

Ornamental aquaculture hit the mainstream news a few years later when it was announced that efforts to rear the blue tang had succeeded. While the yellow tang is a popular fish with aquarium hobbyists, the blue tang had become a celebrity worldwide as the "Dory fish" from the children's film *Finding Nemo* and the sequel *Finding Dory*. The original movie followed the travails of a young clownfish (Nemo), his father (Marlin), and their blue tang sidekick (Dory). The popularity of that film led to a surge in the clownfish market, already a popular hobbyist fish but now an international sensation. But it also popularized the blue tang. On visits to public aquariums, I am often struck by the very few fish that children and their accompanying caregivers can name and how they name them. If you stand near the clownfish or Pacific tanks, you will hear children and their parents exclaim "Nemo!" "Dory!" This popularity has resulted in increased sales but also increased pressure on wild reef ecosystems. The craft for breeding blue tangs came right before the release of *Finding Dory* and was widely reported as an important conservation development.

The University of Florida (UF) Tropical Aquaculture Laboratory closed the cycle on the blue tang. This laboratory is run by fisheries biologists working to develop methods for ornamental aquaculture to strengthen Florida's economic and commercial aquaculture industry. The financial backing for the endeavor came through a combination of federal funding and money from the Rising Tide Foundation. The yellow tang research was partially funded by Rising Tide, and Rising Tide helped link laboratories working on similar aquaculture projects. After the initial successes with yellow tang, Rising Tide paid for UF researchers to visit Hawaii and observe the work in that lab to try to understand how they might transfer that craft into work with blue tangs. The transfer of craft knowledge is most often accomplished by direct viewing, and here we can see how that works in aquaculture. The UF researchers spent time in the OI laboratory watching the methods of yellow tang culture, and with that information and the knowledge of the OI lab, the UF laboratory successfully bred blue tang in captivity.[55]

The development of the captive culture of the Pacific blue tang shows how knowledge of the native environment is combined with aquarium craft and commercial products to create a captive-bred species. The UF researchers state that they worked on a feeding regimen for the blue tang based on knowledge of field observations of prey. According to the authors, copepod nauplii are the preferred first food for marine ornamental fishes because they are the most common food item found in the "gut contents of wild pelagic tropical marine larvae." But they also say that they use them because they have been shown in studies to be the correct size and nutritional content, and that they "stimulate a feeding response in marine fish larvae."[56] Knowledge of the wild habits of blue tang was combined with information about what worked for the yellow tang in Hawaii. But knowledge about wild feeding habits and previous successes in breeding were combined with what food was commercially available. The paper lists the food sources of the captive-bred blue tang, and researchers are combining knowledge of what the fish needs with what will be available to other breeders.

> Pacific blue tang broodstock were fed a varied diet to apparent satiation three to five times daily. The diet consisted of a mixture of a commercially prepared seafood blend (LRS Fertility Frenzy, Larry's Reef Services, Advance, NC, USA); fish eggs (LRS Fish Eggs, Larry's Reef Services); frozen mysis shrimp, *Mysis diluviana* (Piscine Energetics, Inc., Vernon, BC, Canada); and a commercially available 1.7-mm extruded pellet ([EP1—46% crude protein, 16% crude fat, and 2% crude fiber], TDO ChromaBoost, Reed Mariculture, Inc., Campbell, CA, USA).[57]

Throughout the paper, the researchers call attention to the difficulties of working out a diet and feeding regimen that is not only capable of successfully breeding blue tangs but doing so on a large scale. This work requires that the authors acknowledge that hewing closely to natural food sources or habits is not their goal. When discussing further studies about feeding and diet, they state that "further research is needed to investigate different feeding regimes and prey organisms at periods of increased mortalities. The timing of the transition from larvae feeding on copepod nauplii to rotifers will be an important area of research in the future and may impact the potential commercial profitability of this species."[58] Their goal is to create a system that allows consistent breeding of a captive organism that looks like its wild counterpart but is a completely captive creature.

In addition to telling us about shaping a captive species, the breeding of the Pacific blue tang demonstrates the way that knowledge travels through the aquarist network. The researchers at UF worked closely with the OI laboratory in Hawaii. They visited that laboratory and had consistent contact with the aquaculturists working on the yellow tang. It is difficult to trace these interactions in publications, although this community is open when talking about these informal exchanges of information. For instance, a paper by the researchers at UF states that they set up their tanks to be the same as those in Hawaii. This information came from C. K. Callan through "personal communication."[59] In addition, the authors thank the researchers at OI in HPU whose "technical assistance and collaboration were instrumental in this achievement."[60] These two laboratories worked closely together to share tank craft; that sharing is largely done through traveling and physical teaching or informal communications.

The network of users, however, is wider than two universities and a funding agency. This process shows how expansive a community grows when they confront a common problem. Developing captive breeding programs requires many wild-caught subjects for experimentation. The blue tang is a Pacific reef fish and therefore had to be imported to UF at great monetary cost and loss of fish life. This is, of course, anathema to the goals of Rising Tide. During the initial work with blue tang, UF and Rising Tide sought to use existing aquarium specimens to provide usable reproductive materials for experimentation. To do this, Discovery Cove (SeaWorld Orlando), the Birch Aquarium, the Columbus Zoo and Aquarium, and the John G. Shedd Aquarium collected eggs and larvae from tanks containing a variety of ornamental species and sent them to UF for analysis. The lab genetically analyzed them to find out what species had been collected. While it became clear that there were not many viable blue tangs in the group, there were successes that led the group to believe that they could eventually collect and culture quite a few species already in tanks instead of collecting more from the wild.[61]

These nodes in the breeding networks are often ephemeral and difficult to find through literature. The need to trace both the history and current practices of aquaculture in person is why this chapter seems to focus primarily on American, and almost exclusively Floridian, networks. While doing interviews in Central Florida in December 2017, I visited three sites of this network. First, I toured the UF Tropical Aquaculture Laboratory and saw their tanks. Of all the interviews I have done for this book and my general research, it has never been so easy to get information about the parameters of fish keeping and breeding. I was given a tour of the facility and feeding schedules, tank setups, filtration systems, and any other information needed to try my hand at breeding were clearly explained to me. This is in direct contrast to the published materials from the laboratory, which detail the development of methods but are not clear about setting up systems. When asked, it was explained that the job of the laboratory is to disseminate information to the public through interactions and to help the Florida aquaculture industry.

In addition to the link of public aquariums to the UF laboratory, that lab also connected commercial interests back to the laboratory and aquariums. Rising Tide seeks to fund development of aquaculture, but they don't directly fund the commercial culture of those species. Once a process is developed in the laboratory, it must be spread into the commercial community for scaled-up production to disseminate fish to aquariums and hobbyist communities. The UF laboratory is part of the land-grant university system that works with aquaculturists in the state to disseminate knowledge about fish breeding. Their goal is to assist the commercial breeding industry in Florida. As UF develops breeding processes for ornamentals, they pass that information to aquaculturists in Florida to develop large-scale production methods. The laboratory works out the small-scale details of breeding, but the scaling up of production and developing those methods comes from the commercial producers. And when they develop these processes, they often keep that information in the community (and particularly within their own companies). The knowledge that they produce becomes their commodity, and the aquariums and hobbyists purchase captive-bred fish from them.[62]

After visiting that laboratory, I was introduced to a commercial breeder who works directly with the UF laboratory and sells fish to SeaWorld, EPCOT and other aquariums throughout the state. The commercial operation consisted of a set of large, custom-built sheds on private property close to Tampa. While there, I toured his clownfish facility and saw tank after tank of fish being reared for the ornamental market. The breeder was trained at the UF facility and went into commercial work after receiving his bachelor's degree. In addition to the basic clownfish, tanks contained highly valued clownfish varieties that were intended for collectors. Other tanks contained more experimental setups with fish received from the UF laboratory that the breeder was trying to scale up for commercial production. This is another step in the development of aquaculture tank craft: the scaling up from single tank production to consistent production of multiple generations for commercial sale. This breeder might share some of his tank craft with UF so that it can be disseminated throughout the state, but they will also be able to sell those fish directly to SeaWorld and other groups and to maintain some secrecy about their methods.

Finally, I toured Discovery Cove, an artificial reef theme park run by SeaWorld in Orlando, Florida. Discovery Cove is a large lagoon built with artificial coral and live fish to resemble a reef. Visitors don snorkeling gear and swim with a variety of tropical fishes, including sharks. Many of the fish that I saw in that cove were identified as those provided initially to UF for aquaculture development. After the process was developed, the first generation were provided to a breeder, who developed methods for scaling up production. Discovery Cove then purchased those commercially bred fish to stock their lagoon, completing the cycle. Eventually, Discovery Cove would like to collect their fertilized eggs and have those eggs reared and returned as captive-bred fish so that they can completely phase out wild-caught exhibits. While this might seem an expensive endeavor, it would cost more to build their own breeding facilities and stock them with tanks and tinkerers.

This network of academic, commercial, and aquarist communities shows both how knowledge is developed and how each industry is working to make ornamental aquaculture a viable enterprise. In the past, the stops and starts of the research were related to the lack of commercial viability at the end of the process. It is expensive to develop a breeding process, and the outcome is a fish that is significantly more expensive than its wild-caught counterpart. And while captive-bred specimens are often hardier and more likely to survive in captivity (because it is their native environment), hobbyists are still more likely to buy the cheaper, wild-caught specimen.

Today, conservation groups and state agencies have strengthened existing networks and built new avenues by which tank craft can travel. There has been a wealth of new craft produced in the last decade, but much of it becomes stuck in a single section of the network because of political and commercial interests. It isn't the case that there are no publications about this work, but a book on blue tang aquaculture like Hoff's on clownfish—detailing tank setup and parameters, feeding schedules, and every other step for breeding—is not likely anytime soon. This is especially true because ornamental aquaculture is not only of interest for

American communities but also for international groups as well. There are robust ornamental breeding programs throughout the world, especially in China, Taiwan, India, and Italy, to name a few. These programs are succeeding in breeding ornamental species that have not been tried in the United States, and the intricate ties between conservation, industry, and state development appear to be similar in each of these cases.[63]

Understanding Oceans with Oceans under Glass

I have researched a wide array of tank designs for this book. At no time did I encounter such an open group of people willing to talk and be interviewed and followed. This was in direct contradiction to the amount of academic literature about ornamental breeding. While there are quite a few papers on breeding techniques, it is hard to trace the networks of those directly involved and who have a major interest in the subject. The main reason for this seems to be the commercial interests involved.

While academic institutions research breeding, many are attached to the land-grant system and are deeply embedded in their state's economic engines. They are building knowledge to help the breeders in their state achieve economic success. This goal, one that is shared by both the OI and UF laboratories, means that much of their knowledge is passed word-of-mouth and not through academic publications. When I originally emailed UF to get an interview, I thought that it would take months to do so. I've spent a lot of time explaining my work when asking to speak with people about their tank systems: what do I want the information for and how will I use it? But instead UF just asked when I could show up and tour the facility. It was explained to me upon arrival that "this is how it works" and that their job is to help people understand breeding. I won't lie—I almost wanted to start breeding ornamentals just because they made me feel so welcome.

The economic aspect creeps into every corner of ornamental breeding. Commercial breeding communities are necessarily silent in publications because their knowledge is proprietary. While researching, I was unable to access a very large portion of the network in print. Luckily, we have the information from Moe and Hoff and their earliest commercial endeavors, but the highlighting of those authors does not mean they are the only stars in this community. While they are well known (their names came up in every interview I did), they are also known because their publication of information is so rare. Understanding how the marine craft network is built and communicates is especially important when trying to trace knowledge that is spread only partially by formal publication.

Breeding tanks demonstrate the ongoing negotiations of what is important about the natural behaviors and environments in a simulated system. The earliest attempts at ornamental breeding by Hoff and Moe relied on trying to mimic natural environments and then slowly clarifying the essential elements of those environments to basic functions of the organisms. Is the anemone important to clownfish for protection, even if there are no predators? What other support does it provide for the breeding pair, and how can we mimic that in the most minimalist way? Through tinkering, Hoff and Moe found that the anemone was important for a variety of reasons, providing substrate for egg deposit as well as a sense of safety, but that this could be attained using other materials. In other instances, it was found that behavioral patterns deemed unimportant, such as the brooding behaviors of males, were more important to the success of breeding than previously thought. This negotiation shows how the breeding system is both the development of an artificial system and a place to better understand the natural behaviors of reef fishes.

Every species presents new challenges, and one of the most popular species of reef fishes, the wrasse, is a particularly interesting case. Wrasses are protogynous, meaning that they are all born female and one member changes to male to make up a breeding community. During my visit to UF's laboratory, I spoke to a graduate student working on breeding wrasses. The development of the process required the negotiation of variables that tried to simulate their natural environment with limited resources. The researchers used PVC pipe areas to mimic rocks with a partial shade covering because trial and error showed that the fish preferred to catch prey in or around this area. Researchers worked with a variety of substrates on the bottom, originally placing sand from their native environments in the tanks and then substituting a harder, commercially produced aquarium sand when the fish refused to use the native sand for breeding; the artificial sand proved more to their liking. These are the variables that they had worked out, but they still could not get the wrasses to transition consistently to male and the groups to breed successfully.

These negotiations show the basic nature of tinkering in the development of tank craft. The problem could be not enough shade, water quality, too few PVC pipe structures or too many, or possibly too many or too few fish in the same tank. It could be that there is too much light or too little, or it could be the wrong UV spectrum. It could be that wrasses require a spike or dip in temperature or a shift in water chemistry to induce spawning or to regulate the reproduction in both males and females. Or it could be that the fishes in the laboratory aren't mature enough to reproduce even though they reproduce at that age in the wild. Because the captive environment is different from the wild, the natural behavior of a species is not always an indication of how they will develop and behave in captivity. But every time they have a success in the step taken, every piece of tank craft tells the aquaculturists something about their fish and can be passed to field researchers to see whether those variables are also important in the wild. Oceans under glass, combined with fieldwork, can tell us much about the larger marine world.

This negotiation, the transfer of wild into captive, will be especially important if conservationists ever hope to use aquarium technology to replenish natural environments currently being destroyed. Knowing how and why animals make certain decisions and what environmental changes they will tolerate and thrive in versus those that will be rejected will become important as humans continue to shape future oceans. Right now, the oceans are being shaped accidentally, but the bioengineering of reefs and other spaces suggest that we are not far from purposely building reefs to maintain communities of organisms that would otherwise be lost without our interventions. The knowledge produced in these tanks will shape those human interventions.

Conclusion

"You Are the Ocean"

Scaling Up Oceans under Glass

It's a blustery day in January 2017, and I'm standing next to a large, replicated ocean complete with a beach and stocked with a wide variety of fish and sunburned tourists. Discovery Cove is a theme park owned and operated by SeaWorld that caters to individuals who want to see submarine wildlife up close but don't want to leave the inland comforts of Orlando, Florida. Tourists can take a day off from riding Space Mountain and touring the Wizarding World to don a wetsuit and immerse themselves in a large aquarium tank built to resemble a coral reef.

Outside the water, it is a cold enough day that I'm wearing a coat, but wetsuit clad families float lazily in the water, occasionally kicking to follow schools of fish around the shallow lagoon. In an adjacent area, you can swim with sharks. Aquarists keep a sharp eye out for divers in distress, answer questions about the animals inhabiting the tank, and make reports on the general health, feeding, and behaviors of the animal residents. Most of these aquarists are currently majoring in marine science or have recently graduated from local universities; they are hoping to break into the marine science world through this work.[1]

There are several types of visitors to Discovery Cove. The first is the uninitiated marine visitor. By this I mean that many of the visitors are individuals who come from an inland location and who might have never snorkeled, dived, or even swum in the ocean. They might be individuals for whom swimming in open water is foreign or even scary (there were a lot of flotation devices in use on the day that I visited). This is possibly the closest they will ever get to exploring a coral reef or seeing a submarine environment in situ. Diving in a large, man-made lagoon decreases the stress on the visitors *and* on the wild coral reefs that would be viewed by these tourists.

The other set of visitors are those who have experienced ocean dives but seek out the Discovery Cove experience for other reasons. When I began snorkeling as a child in Florida, I often lamented the lack of fish in

FIGURE 6.1 Discovery Cove, SeaWorld, Orlando, Florida. The tank represents not just the future of public aquarium exhibit but also possibly the future of the wider oceans. Author's collection.

the waters just off the beach. To find more colorful fishes and more inhabited waters, my family would travel to beaches with man-made coral reefs. These reefs effectively consolidated populations into spaces more easily accessible to divers from the beach. But those reefs still contain fewer varieties and total numbers of fish than Discovery Cove. In fact, there isn't anywhere in the world where you will see the combination of organisms in such quantities for uninterrupted study as this tank. I was initially surprised when my guide told me that quite a few experienced divers and snorkelers go to Discovery Cove, but as I thought about it, it seemed like a good reason to go: if your goal is to float with a bunch of fish on a beautiful coral reef and you want to see as many fish as possible, this seems like a good alternative to the wild.[2]

It is not a stretch to apply Baudrillard's theories about the simulations to Discovery Cove. According to Baudrillard, simulations that become too real in turn become that which they seek to simulate. His examples often involve spaces such as Disney, a theme park that simulates a type of reality (happiness and childhood) and over time *becomes* this reality to individuals. What was once a place to "make believe" becomes a real space that is woven into the fabric of reality and replaces the real. Disney, which previously was hyperreal, becomes the new real, and childhood happiness is this hyperreality.

Discovery Cove can be read in this light. It isn't the ocean, but instead a simulation of the ocean. For those who visit the Cove instead of having ever experienced the marine environment, it becomes the ocean in their imagination. And the memory of that space is what they will recall when they think of reefs and their physical experience with them. In addition, if they do dive in the marine world, they might experience it as Beebe described; it will become less real to them than the hyperreality of the Cove. If they cannot shake themselves out of that simulation, the "real" marine world will be supplanted by the simulation. For those who know the marine environment already, the hyperreality of the Cove might seem less of an all-encompassing simulation, one that cannot become reality. But this, too, is not impossible. Choosing simulation over reality means that there is something more enticing about that experience and shows that there is something in that simulation that is more appealing as a reality. The simulation is embraced, and the reality rejected, allowing the simulation to become the new real environment.[3]

But unlike many of Baudrillard's examples, in which simulations become reality but the original reality still exists, tanks such as Discovery Cove might actually supplant the marine realm in the future.

In the last decade the outlook for coral reefs, and marine environments more generally, has become dire. As climate change and human-encroachment issues have become seemingly irreversible, conservationists have turned from the notion of "saving" the oceans that we have to building better and more resilient marine worlds. This focus on bioengineering has brought places such as Discovery Cove and the tank craft used to create them into greater focus. The lagoon isn't just stocked with fish, it is currently stocked with as many captive-bred fish species as possible. Because of the relationship between SeaWorld, Rising Tide, and the laboratories and commercial enterprises breeding marine fish in Florida and around the world, the Cove contains schools of captive-bred fish. And while the reef structure is mostly concrete now, it is only a matter of time before Discovery Cove introduces hardier species of coral to complete the snorkeling experience. And those corals will be captive bred as well.

Discovery Cove may well be the future ocean, full of the animals that humans are capable of breeding, rearing, and maintaining.

Tank craft is integral to building a resilient ocean capable of weathering climate change. This book has shown how over the course of the last century a network of tank users boxed the open ocean. This section will show how that network, imbued with knowledge of impending climate disaster, are seeking to scale up those boxes. Their hope is that the craft developed in oceans under glass will save oceans outside the glass.

Building Better Oceans: Coral Reefs

The degeneration of the world's oceans has prompted new conservation research over the last thirty years. In 1995 Baruch Rinkevich suggested breaking coral into smaller pieces (known as *fragging*) and isolating those fragments in underwater nurseries to be gardened for replacement into the larger reef system. Rinkevich felt that coral gardening could help restore depleted reef systems. This form of active restoration of existing reefs is practiced worldwide.[4] But by 2015 the coral crisis deepened, and more drastic measures were proposed. Madeleine van Oppen and Ruth Gates suggested that coral that showed resilience to bleaching and disease be brought into the laboratory and genetically analyzed. After gaining an understanding of the mechanisms that produce resilience in those coral, they could be bred and the stronger corals replaced on the reefs.[5] Today, coral conservation and restoration involves a combination of these strategies. But regardless of the methods employed to restore reefs, oceans under glass have an integral role to play.

The first step to building a resilient coral reef is to understand the threats to them and how corals react. Corals in captivity can tell us much about the threats to wild reefs, and there are many ways that tanks can contribute to this understanding. One way that coral tanks can contribute to knowledge of the disasters facing the oceans occurs during the process of trying to maintain these systems. When I was visiting in the summer of 2017, the marine tank at the National Museum of Natural History (NMNH) that replaced Adey's in the Marine Hall was struggling with a mysterious disease. This tank educates visitors on marine ecosystems; a commercial enterprise maintains and stocks the tank with the help of an automated system, daily checks by docents, and monthly visits from a vet. But the tank had been struggling with a disease that was attacking the coral. The disease looked like the white or black band disease decimating coral reefs throughout the Florida Keys, but it was a mystery to those who were keeping the tank. This was the second major coral die-off in the tank, and this newest disease had a different symptomology and presentation than the previous issue. I was allowed to look behind the scenes with a group trying to both save the tank and understand what could be going wrong.

FIGURE 6.2 Researcher reaching into the National Museum of Natural History tank to cut affected sections of coral from the main group in hopes of saving the larger coral and to run tests to better understand the disease. Author's personal photo, June 26, 2017.

The group consisted of three different members of the marine network: Mark, a commercial aquarist with Reef Escape in charge of the maintenance of the tank; Brittany, a graduate student at a local university, previous volunteer at NMNH, and employee of Reef Escape; and Brent, an aquatic veterinarian who oversaw the health of the coral and fish in the tank. Initially, the group performed the regular checkup and monthly maintenance on the tank. But afterward, they began to discuss the mystery of the disease. Each spoke about their experiences with the tank and what they felt might be done to save the coral. A particular focus was the relationship between the coral disintegration and the fish communities. Was the water chemistry shifting, and if so, how was it affecting other inhabitants of the tank? Both Mark and Brent agreed that the fish were behaving normally and were preparing to mate based on observed behaviors, suggesting that they were not under new stress from the coral disease.[6]

After some consideration, it was agreed that the diseased sections could be trimmed to help maintain as much of the healthy coral as possible and to make room for new coral to be placed in the tank, and the diseased pieces could be preserved and transferred to the laboratory for analysis. To maintain the aesthetics of the tank, more coral would be added. Brittany suggested sampling each new coral to run genetic tests before they were added. Mark trimmed off pieces of the affected coral, and Brittany transferred them to the freezer before transporting them to the laboratory to be looked at with an electron microscope in the hopes of identifying and better understanding the cause of the disease. The group did other maintenance work, finished their paperwork, and left.

This exchange shows how tanks can become spaces for understanding current marine issues and tackling them. The diseases that occur in this tank could be introduced from a variety of places—new fish or coral added to the tank, food sources, or even water additives. But the likelihood is that if it is occurring in this tank, it is something that is also occurring in the open ocean. This tank is not a laboratory system; it is not built to be an experimental tank. There is tension in the interactions when trying to balance the science and aesthetics of the tank. The tank is supposed to be aesthetically pleasing to visitors of the Marine Hall, so adding new corals as soon as possible will help the tank look fuller and more robust. However, adding more corals without understanding what is happening within the tank could both kill those corals and potentially hurt the understanding of the disease's life course and effect on the original coral and fishes. But these tensions do not completely negate the usefulness of the tank for knowledge creation. Each member of the network who helps maintain the tank will produce knowledge from this exchange. The combination of their history with the tank and their individual expertise will lead to recognition of the disease symptoms in other tanks and possibly a solution to that problem when it arises. But there will also be knowledge produced by the laboratory work. These differing forms of knowledge will circulate throughout the network via tank craft and eventually contribute to more understanding of this disease.

In addition to informal networks and nonexperimental tanks, academic biologists have begun using tanks to understand more clearly these disease threats to wild populations. In a more controlled setting, Valerie Paul is researching coral diseases in a group of open-air tanks at the Smithsonian Marine Station in Fort Pierce, Florida. Just across the street from the St. Lucie County Aquarium, Paul and her lab have set up an outdoor laboratory of tanks, each containing coral frags.[7] The tanks were set up to test disease transmission of stony coral tissue loss disease (SCTLD), a disease currently affecting large swaths of reef throughout the Florida Keys. Tanks contained several species of frags, the same species of frags with different levels of the same disease, and a combination of open and closed circulation to test how the disease spreads and how fast. In a paper published in November 2019, Paul and colleagues found that

> Aquaria studies showed disease transmission occurred through direct contact and through the water column for corals from both regions. However, transmission success was higher for corals with acute vs. subacute lesions. There was 100% transmission for both test species, *M. cavernosa* and *Meandrina meandrites*, touching acute lesions. Among the three species touching subacute lesions, the disease transmitted readily to *Orbicella faveolata* (100%) followed by *M. cavernosa* (30%) with no transmission occurring with *Porites astreoides*. Diseased fragments of all species tested responded to antibiotic treatment with a cessation or slowing of the disease lesions suggesting that bacteria are involved in disease progression.[8]

The public aquarium and laboratory tank systems described here will make it possible to better understand and conserve coral for the future. But understanding threats to coral is only the first step: the next is to remove them from the open ocean for safe keeping and propagation.

In December 2019, following the hottest start to Australia's summer on record, it was announced that a conservation group, the Great Barrier Reef Legacy group, would be starting a Coral Biobank Project. This project seeks to preserve as much coral diversity from the Great Barrier Reef as possible by storing and rearing coral in a holding facility in Port Douglas, Australia. It will also send frags to advanced hobbyists and public aquariums worldwide in the hopes that some of this diversity will survive and thrive in captivity. In the event that the marine environment stabilizes, it might be possible to replant these frags and rebuild the Great Barrier Reef. The ultimate goal of projects like this is the planting of genetically diverse corals in struggling oceans. Conservation groups such as SECORE (Sexual Coral Reproduction) work to breed corals in captivity to increase genetic diversity before they transplant them in wild reefs. Their work is similar to other organizations such as Hope for Reefs and the Reef Resilience Network. All of these groups rely on the knowledge of those using tanks to understand how to propagate and transplant corals.[9]

But the scaling up of oceans under glass can leave behind those organisms that cannot be transferred into tanks in the first place. Tank craft is difficult, and many marine organisms cannot survive and thrive in them. This is not indicative of their importance to ecosystems, but it might foretell their deletion from future oceans. For instance, one of the most important animals on the Atlantic reef tract is the long-spined sea urchin *Diadema antillarum*. A collapse of this community with over 93 percent mortality led to a collapse of reefs in and around the Caribbean in the

1980s.[10] It is only in 2020 that the Florida Aquarium and the University of Florida, working with Martin Moe, succeeded in breeding and releasing *Diadema antillarum*.[11]

Great white sharks are one of the most recognizable marine species, yet they offer extreme difficulties in captivity. No aquarium has maintained the species for the long term, with only a few larger public aquariums able to keep juveniles for a short time.[12] In addition to species that cannot be cultivated, certain ecosystems are difficult to reproduce in tanks. The midnight zone, the deepest part of the ocean, is a space of extreme pressure and little light. To study organisms from this zone in captivity, the Monterey Bay Aquarium works under red light or in the dark, with images recorded via low-light video cameras. The aquarium is currently developing craft to maintain some organisms from this zone in captivity, but the ability to breed and replace them is far off.[13]

There are many more species that we have not yet tried to bring into the laboratory, and the development of the craft knowledge needed to understand them and keep them alive in captivity will take decades or even centuries. But the time for slowly developing tank craft to maintain oceans under glass is running out. Instead, as we look toward the need to bioengineer the ocean for survival of the oncoming climate disaster, it might be the case that the only reefs that survive are those that have been effectively grown and maintained in captivity to date. The future oceans are being shaped by two challenges: climate change and current tank craft.

In this way, we can see how Discovery Cove might indeed be a simulation that supplants reality. The Cove is stocked with fish bred in captivity, and the sharks that you can pay to swim with will be bred in captivity. Over time, it isn't impossible to imagine that the cement reef will be replaced with more and more captive-bred coral species to give visitors a more "authentic" experience and to protect those corals from certain death in the real ocean. Eventually, Discovery Cove, and oceans under glass, will be stocked with the only surviving and survivable ocean inhabitants.

The work done in tanks today is building the ocean of tomorrow. Tank crafters are shaping our future oceans with every success or failure. As Mary Akers whispers, they "are the ocean."

Acknowledgments

Writing this book would not have been possible, or nearly as enjoyable, without the amazing community of aquarists who have taken the time to share their tanks and knowledge with me. Some of the earliest interviews performed for this book took place in 2011, and over the course of a decade I've visited a wide range of laboratories, public aquariums, and homes containing aquariums. Unfortunately, it would take far too long to list all these places and people separately, but know that if you spoke with me, whether your name appears in this book or just in my notes, your kindness and enthusiasm for your work has earned my eternal gratitude. All mistakes or misunderstandings are my own, and I apologize in advance that this book could not contain every name of every important person in all the fields I study. If there is one lesson I hope the readers of this book come away with, it is that the aquarium craft community is full of amazing people doing wonderfully interesting stuff.

Much of the archival work and fieldwork for this book was conducted during predoctoral (2011–2012) and postdoctoral (2017) fellowships from the Smithsonian Institutional Archives. Pam Henson has always been an important mentor who maybe understood my project (and its historical holes) long before I did. Talking with her has always pushed me back on track and farther down the road. Nancy Knowlton, Bill Hoffman, and everyone that I interviewed and shadowed in DC and at the aquarium and laboratory in Fort Pierce, Florida, made this book much better than it would have been without their help.

Jane Maienschein was kind enough to invite me to participate in workshops during summers at the Marine Biological Laboratory (MBL). Several of this book's arguments (and the seeds of chap. 5) were sketched out during these workshops during talks and drinks with colleagues. In addition, I was able to interview a wide range of scholars in Woods Hole, including the engineering and aquarist staff of the MBL. Finally, Jen

Walton, the aquarist of the MBL / Woods Hole Oceanographic Institute library has helped me find a million resources in record time.

My time at Florida State University and the University of Pennsylvania set me up with the best resources in the world: an appreciation of interdisciplinarity and connections with a wide network of scholars who continually push me to think broadly about my work. Jessica Martucci's guidance has been invaluable. Kristoffer Whitney, Emily Pawley, Peter Collopy, and Deanna Day have all been fantastic sounding boards, offering both academic conversation and career advice. Susan Lindee, Mark Adams, John Tresch, Michael Ruse, and Frederick Davis gave me great intellectual tools to tinker with and only tried to rein me in occasionally.

Helen Rozwadowski and Katharine Anderson have shown great enthusiasm for this project. Without them, this book would be a half-finished pipe dream always listed on my CV as forthcoming. In addition to this project, they have worked to bring together the marine history community, and through them I have found some of my favorite people to talk to, including Antony Adler and Penelope Hardy. I couldn't write this book without the strong marine history scholars who have commented on drafts, chapters, and general musings.

The administration and my colleagues at the Stevens Institute of Technology have been helpful in the final stages of this manuscript. The entire College of Arts and Letters and my students have showed enthusiasm about my work. My undergraduate research assistant Wiktoria Jurgowski's work on the bibliography got this book across the finish line.

Finally, my family and friends have been living with this book for a very long time. They have been cheerleaders through the publishing process even as they might have started to wonder whether I would ever just hurry up and finish it. My sister Jeannie Knotts has always believed in my projects, even when I couldn't articulate them clearly. Sabrina Jamil, Charis Stiles, Erikka Bahnuk, and the rest of my friends and family are always willing to listen to me talk about fish. My mother, Tammie Harmon, and my father, Gary Muka, let me wander widely and encouraged all my far-fetched notions. My husband, Chris Zarpentine, has never doubted my work, and without him it would have been an even longer process. He uprooted our family for my postdoc, drove me to interviews while I was heavily pregnant, and fell asleep many nights surrounded by books on aquariums, listening to me try to explain my ideas a little more clearly. He has always been an example of hard work and commitment to academic pursuits, but it is his devotion to our family that has allowed me to complete this project. It is to him that I dedicate this book.

Notes

Preface

1 “How Much of the Ocean Have We Explored?,” National Ocean Service, NOAA, January 1, 2009, https://oceanservice.noaa.gov/facts/exploration.html.

2 “A Decade of Discovery,” Census of Marine Life, 2010, http://www.coml.org/; see also Niki Vermeulen *Supersizing Science: On Building Large-scale Research Projects in Biology* (Boca Raton, FL: Dissertation.com, 2010), chap. 3.

3 D. G. Lopukhin, “Aquaculture of Ancient Rome and Ornamental Fishkeeping,” in *Дни науки* (2019): 215–19; Victor D. Thompson, William H. Marquardt, Michael Savarese, Karen J. Walker, Lee A. Newsom, Isabelle Lulewicz, Nathan R. Lawres, Amanda D. Roberts Thompson, Allan R. Bacon, and Christoph A. Walser, “Ancient Engineering of Fish Capture and Storage in Southwest Florida,” *Proceedings of the National Academy of Sciences* 117, no. 15 (2020): 8374–81; James Arnold Higginbotham, *Piscinae: Artificial Fishponds in Roman Italy* (Chapel Hill: University of North Carolina Press, 1997.

4 David Allen, “23 Tastes and Crazes,” *Cultures of Natural History* (1996): 394.

5 Bruno Latour, *Science in Action: How to Follow Scientists and Engineers through Society* (Cambridge, MA: Harvard University Press, 1987).

6 Dolly Jørgensen, Finn Arne Jørgensen, and Sara B. Pritchard, eds., *New Natures: Joining Environmental History with Science and Technology Studies* (Pittsburgh, PA: University of Pittsburgh Press, 2013).

7 Luis A. Campos, Michael R. Dietrich, Tiago Saraiva, and Christian C. Young, eds., *Nature Remade: Engineering Life, Envisioning Worlds* (Chicago: University of Chicago Press, 2021).

8 Samantha Muka, “Historiography of Marine Biology,” in *Handbook of the Historiography of Biology*, ed. Michael R. Dietrich, Mark E. Borrello, and Oren Harman (Cham, Switzerland: Springer, 2021): 435–59.

9 Walter Adey and Karen Loveland, Box 10, Coral Reefs, Roll #135, Marine System Lab, June 7, 1982, 16 mm, Smithsonian Institution Archives, Accession 05–263, Smithsonian Productions.

Chapter One

1 "EarthWaterFireStudio," Etsy, https://www.etsy.com/shop/EarthWater FireStudio?ref=seller-platform-mcnav; Samantha Edmonds, "The Miraculous Journey of a Captive-Bred Hermit Crab," *Outline*, October. 18, 2019, https:// theoutline.com/post/8116/hermit-crabs-breeding-captivity?zd=1&zi= 6qeiqbye.

2 Mary Akers, "Summer 2018 Breeding Attempt," https://maryakers.com /summer-2018-breeding/.

3 Mary Akers, "Summer 2017 Breeding Attempt," September 1, 2017, https:// maryakers.com/summer-2017-breeding-attempt/.

4 Akers, "Summer 2018 Breeding Attempt," November 13, 2018.

5 JadeWolf42, March 11, 2019, https://www.etsy.com/listing/578107906/pea -pod-dish-in-emerald-green?show_sold_out_detail=1&ref=nla_listing _details.

6 Edmonds, "Miraculous Journey."

7 Lynn K. Nyhart, *Modern Nature: The Rise of the Biological Perspective in Germany* (Chicago: University of Chicago Press, 2009).

8 Ciemon Frank Caballes and Morgan S. Pratchett, "Environmental and Biological Cues for Spawning in the Crown-of-Thorns Starfish," *PLoS* 12, no. 3, (2017): e0173964, https://www.ncbi.nlm.nih.gov/pmc/articles/PMC 5371309/.

9 Claude Lévi-Strauss, *The Savage Mind* (Chicago: University of Chicago Press, 1966), 17.

10 François Jacob, "Evolution and Tinkering," *Science* 196, no. 4295 (1977): 1164.

11 Kathleen Franz, *Tinkering: Consumers Reinvent the Early Automobile* (Philadelphia: University of Pennsylvania Press, 2011), 11.

12 Caitlin Donahue Wylie, "'The Artist's Piece Is Already in the Stone': Constructing Creativity in Paleontology Laboratories," *Social Studies of Science* 45, no. 1 (2014): 3.

13 Ibid., 2.

14 Franz, *Tinkering*, 9.

15 Bess Williamson, "Electric Moms and Quad Drivers: People with Disabilities Buying, Making and Using Technology in Postwar America," in *Disability, Space, Architecture: A Reader*, ed. Jos Boys, 198–306 (London: Routledge, 2017); Bess Williamson, *Accessible America: A History of Disability and Design*, 2 vols. (New York: New York University Press, 2019); Jeremy A. Greene,

"Do-It-Yourself Medical Devices: Technology and Empowerment in American Health Care," *New England Journal of Medicine* 374, no. 4 (2016): 305–8.

16 Frank Nutch, "Gadgets, Gizmos, and Instruments: Science for the Tinkering," *Science, Technology, & Human Values* 21, no. 2 (1996): 214–28.

17 Interview, University of North Carolina, Wilmington, September 24, 2015.

18 Susanne K. Schmidt, Raymund Werle, K. Susanne, and Trevor Pinch, *Coordinating Technology: Studies in the International Standardization of Telecommunications* (Cambridge, MA: MIT Press, 1998), 4.

19 Joan H. Fujimura, "Standardizing Practices: A Socio-History of Experimental Systems in Classical Genetic and Virological Cancer Research, ca. 1920–1978," *History and Philosophy of the Life Sciences* (1996): 3–54.

20 Franz, *Tinkering.*

21 Kathryn Packer and Andrew Webster, "Patenting Culture in Science: Reinventing the Scientific Wheel of Credibility," *Science, Technology, & Human Values* 21, no. 4 (1996): 427–53; Paul Lucier, "Court and Controversy: Patenting Science in the Nineteenth Century," *British Journal for the History of Science* 29, no. 2 (1996): 139–54.

22 Collins, "Tacit Knowledge," 72.

23 Ibid. Harry M. Collins, "What Is Tacit Knowledge?," in *The Practice Turn in Contemporary Theory*, ed. Theodore R. Schatzki (London: Routledge, 2005), 115–28.

24 Harry Collins and Robert Evans, *Rethinking Expertise* (Chicago: University of Chicago Press, 2008).

25 For information on citizen science see Gwen Ottinger, "Buckets of Resistance: Standards and the Effectiveness of Citizen Science," *Science, Technology, & Human Values* 35, no. 2 (2010): 244–70; Aya H. Kimura and Abby Kinchy, "Citizen Science: Probing the Virtues and Contexts of Participatory Research," *Engaging Science, Technology, and Society* 2 (2016): 331–61. For conversations on amateur science, see W. Patrick McCray, "Amateur Scientists, the International Geophysical Year, and the Ambitions of Fred Whipple," *Isis* 97, no. 4 (2006): 634–58; Elizabeth Keeney, *The Botanizers: Amateur Scientists in Nineteenth-Century America* (Chapel Hill: University of North Carolina Press, 1992); Rebecca J. McLain, Harriet H. Christensen, and Margaret A. Shannon, "When Amateurs Are the Experts: Amateur Mycologists and Wild Mushroom Politics in the Pacific Northwest, USA," *Society & Natural Resources* 11, no. 6 (1998): 615–26. For resources on "ways of knowing," see John V. Pickstone, *Ways of Knowing: A New History of Science, Technology, and Medicine* (Chicago: University of Chicago Press, 2001). See also the special issue of *History of Science* 49, no. 3 (2011).

26 Helen M. Rozwadowski, *Vast Expanses: A History of the Oceans* (London: Reaktion Books, 2018), chap. 7.

27 Karen Knorr-Cetina, *Epistemic Cultures: How Science Makes Sense* (N.p.: n.p., 1995); Bruno Latour, *Science in Action: How to Follow Scientists and Engineers through Society* (Cambridge, MA: Harvard University Press, 1987); Sharon Traweek, *Beamtimes and Lifetimes* (Cambridge, MA: Harvard University Press, 2009).

28 There are volunteers that perform some maintenance functions at public aquariums. I do not label these individuals as public aquarists but instead include them in the hobbyist community. There are no individuals who fall into this category in this book, but the history of aquarium volunteers is very interesting, and many of the professionals in this community started out as volunteers. It is important to acknowledge that this is a very important community that produces knowledge about tanks and tinkering.

29 Philip F. Rehbock, "The Victorian Aquarium in Ecological and Social Perspective," in *Oceanography: The Past*, ed. M. Sears and D. Merriman, New York: Springer, 1980), 55239; Mareike Vennen, *Das Aquarium: Praktiken, Techniken und Medien der Wissensproduktion (1840–1910)* (Göttingen: Wallstein, 2018).

30 Vernon N. Kisling, *Zoo and Aquarium History: Ancient Animal Collections to Zoological Gardens* (Boca Raton: CRC, 2000); Ben A. Minteer, Jane Maienschein, and James P. Collins, eds., *The Ark and Beyond: The Evolution of Aquarium and Zoo Conservation* (Chicago: University of Chicago Press, 2018).

31 Samantha Muka, "Conservation Constellations: Aquariums in Aquatic Conservation Networks," in Minteer Maienschein, and Collins, *The Ark and Beyond*.

32 Ibid.

33 Ibid.

34 Vernon Kisling. "Historic and Cultural Foundations of Zoo Conservation: A Narrative Timeline," in Minteer Maienschein, and Collins, *The Ark and Beyond*, 41–50.

35 *Drum and Croaker: A Highly Irregular Journal for the Public Aquarist*, http://drumandcroaker.org/.

36 For information on the Monterey Bay Aquarium Jelly School, see the Monterey Bay Aquarium Blog, Monterey Bay Aquarium, "Behind the Scenes at Jelly School," https://montereybayaquarium.tumblr.com/post/143489586643/behind-the-scenes-at-jelly-school.

37 Piscine Energetics, "Coral Double Header: Water Parameters and Project Coral Lab," Zoom meeting, April 16, 2020. This meeting was one of many that occurred between March and July, 2020. Other communities, including the Reef Resilience Network, hosted online workshops during this time as well.

38 Samantha K. Muka, "The Right Tool and the Right Place for the Job: The Importance of the Field in Experimental Neurophysiology, 1880–1945,"

History and Philosophy of the Life Sciences 38, no. 3 (2016): 7; "Illuminating Animal Behavior: The Impact of Malleable Marine Stations on Tropism Research," in *From the Beach to the Bench: Why Marine Biological Studies?*, ed. Jane Maienschein, Karl Matlin, and Rachel Ankeny (Chicago: University of Chicago Press, 2020), 119–43.

39 The exception to this is the use of tanks in fisheries biology. This academic community, talked about in chapter 5, is much more open about their use of tanks for research and are more likely to speak about the parameters of their tank systems in publications. However, much of their interchange of tank craft comes in the form of face-to-face explanations through extension programs and government groups. American Fisheries Society, https://education.fisheries.org/education-links/2015-revised-and-updated-master-list-of-fisheries-programs/ (accessed December 19, 2019; no longer posted). For information on the rise of fisheries research throughout the world, see J. M. Hubbard, *A Science on the Scales: The Rise of Canadian Atlantic Fisheries Biology, 1898–1939*. Toronto: University of Toronto Press, 2006); Tim D. Smith, *Scaling Fisheries: The Science of Measuring the Effects of Fishing, 1855–1955* (Cambridge: Cambridge University Press, 1994); Carmel Finley, *All the Fish in the Sea: Maximum Sustainable Yield and the Failure of Fisheries Management* (Chicago: University of Chicago Press, 2011).

40 Silvia Granata, "'Let Us Hasten to the Beach': Victorian Tourism and Seaside Collecting," *Lit: Literature Interpretation Theory* 27, no. 2 (2016): 91–110; Philip F. Rehbock, "The Victorian aquarium in ecological and social perspective," in *Oceanography: The Past*, ed. M. Sears and D. Merriman (New York: Springer, 1980), 522–39; Bernd Brunner, *The Ocean at Home: An Illustrated History of the Aquarium* (London: Reaktion Books, 2012); Smith, Jonathan. "Eden under Water: The Visual Natural Theology of Philip Gosse's Aquarium Books," paper presented at the conference on "Nineteenth-Century Religion and the Fragmentation of Culture in Europe and America," Lancaster, England, July 1997; Allen, "23 Tastes and crazes," 394.

41 Tomoyoshi Komiyama, Hiroyuki Kobayashi, Yoshio Tateno, Hidetoshi Inoko, Takashi Gojobori, and Kazuho Ikeo, "An Evolutionary Origin and Selection Process of Goldfish," *Gene* 430, no. 1/2 (2009): 5–11.

42 Hugh McCormick Smith *Japanese Goldfish, Their Varieties and Cultivation: A Practical Guide to the Japanese Methods of Goldfish Culture for Amateurs and Professionals* (Washington, DC: W. F., 1909); Charles Nash Page, *Aquaria: A Treatise on the Food, Breeding, and Care of Fancy Goldfish, Paradise Fish, Etc.* (Des Moines, IA: published by the author, 1898).

43 Harriet Ritvo, "Pride and Pedigree: The Evolution of the Victorian Dog Fancy," *Victorian Studies* 29, no. 2 (1986): 227–53, and *The Animal Estate: The English and Other Creatures in the Victorian Age* (Cambridge, MA: Harvard University Press, 1987); Emily Pawley, "The Point of Perfection: Cattle

Portraiture, Bloodlines, and the Meaning of Breeding, 1760–1860," *Journal of the Early Republic* 36, no. 1 (2016): 37–72.

44 Judith Hamera, *Parlor Ponds: The Cultural Work of the American Home Aquarium, 1850–1970* (Ann Arbor: University of Michigan Press, 2012).

45 Federation of American Aquarium Societies, https://www.faas.info/.

46 George Sprague Myers to William Innes, May 16, 1932, Box 3, George Sprague Myers folder, William T. Innes Papers, American Philosophical Society, Philadelphia.

47 Susan Leigh Star and James R. Griesemer, "Institutional Ecology, Translations, and Boundary Objects: Amateurs and Professionals in Berkeley's Museum of Vertebrate Zoology, 1907–39," *Social Studies of Science* 19, no. 3 (1989): 393.

48 Martin A. Moe, *Breeding the Orchid Dottyback,* Pseudochromis fridmani: *An Aquarist's Journal* (Plantation, FL: Green Turtle, 1997); "A Message from Martin Moe," Rising Tide Conservation (blog), https://www.risingtideconservation.org/a-message-from-martin-moe/; personal communication, June 2019; http://macnaconference.org/2018/talks/martin-moe/ (accessed December1, 2011; no longer posted).

49 William Beebe, *Half Mile Down* (New York: Duell, Sloan and Pearce, 1951), 77.

50 It is tempting to refer to aquariums as models. They resemble environmental models written about by Christine Keiner, David Munns, and Mott Green. Green refers to more simplistic climate models as "toy models." However, the modeling literature, especially that literature in the sociology of science, has focused primarily on quantitative modeling and strict control of variables in laboratory environments. The aquariums I speak about occur both inside and outside aquariums, and the control and attention to detail does not fall into these specifications. Therefore, I have opted to call them *replications* and *engineered environments* instead. For a historical analysis of ecosystem and climate modeling, see Christine Keiner, "Modeling Neptune's Garden: The Chesapeake Bay Hydraulic Model, 1965–1984," in *The Machine in Neptune's Garden: Historical Perspectives on Technology and the Marine Environment,* ed. Helen M. Rozwadowski and David K. Van Keuren (Sagamore Beach, MA: Science History, 2004), 273–314; David P. D. Munns, *Engineering the Environment: Phytotrons and the Quest for Climate Control in the Cold War* (Pittsburgh, PA: University of Pittsburgh Press, 2017); Silvia Granata, "'At Once Pet, Ornament, and "Subject for Dissection"': The Unstable Status of Marine Animals in Victorian Aquaria," *Cahiers victoriens et édouardiens* 88 (Autumn 2018); Mott Greene, "Arctic Sea Ice, Oceanography, and Climate Models," in *Extremes: Oceanography's Adventures at the Poles*, ed. Keith R. Benson and Helen M. Rozwadowski (Sagamore Beach, MA: Science History, 2007), 305–12.

51 Etienne Benson discusses replicate environments in his book *Surroundings*. While I am using the word *environment* as relatively unproblematic in this

book, Benson suggests that it is actually a slippery word. Etienne S. Benson, *Surroundings: A History of Environments and Environmentalisms* (Chicago: University of Chicago Press, 2020), 32. For a discussion of engineered environments and the role of engineering in these environments in twentieth-century biology, see Luis A. Campos, Michael R. Dietrich, Tiago Saraiva, and Christian C. Young, eds., *Nature Remade: Engineering Life, Envisioning Worlds* (Chicago: University of Chicago Press, 2021).

52 Benson, *Surroundings*; Munns, *Engineering the Environment.*

53 John Allen and Mark Nelson, "Overview and Design Biospherics and Biosphere 2, Mission One (1991–1993)," *Ecological Engineering* 13, nos. 1–4 (1999): 15–29.

54 Susanne Bauer, Martina Schlünder, and Maria Rentetzi, *Boxes: A Field Guide* (Manchester: Mattering Press, 2020), 41, https://www.matteringpress.org/books/boxes.

55 Nyhart, *Modern Nature*, 252.

56 Alain Corbin, *The Lure of the Sea: The Discovery of the Seaside in the Western World, 1750–1840* (Berkeley: University of California Press, 1994); Margaret Cohen, *The Novel and the Sea* (Princeton, NJ: Princeton University Press, 2021).

57 Ibid., 71.

58 Christopher Hamlin, "Robert Warington and the Moral Economy of the Aquarium," *Journal of the History of Biology* (1986): 131–53; Aileen Fyfe, *Science and Salvation: Evangelical Popular Science Publishing in Victorian Britain* (Chicago: University of Chicago Press, 2004).

59 Hamera, *Parlor Ponds.*

60 Granata, Silvia. "The Dark Side of the Tank: The Marine Aquarium in the Victorian Home." In *Paraphernalia! Victorian Objects*, ed. Helen Kingstone and Kate Lister (London: Routledge, 2018), 81–98, and "The Victorian Aquarium as a Miniature Sea," *Underwater Worlds: Submerged Visions in Science and Culture* 19, no. 1 (2019): 108; Rebecca Duffy, "The Age of Aquaria: The Aquarium Pursuit and Personal Fish-Keeping, 1850–1920" (PhD diss., University of Delaware, 2018).

61 Dolly Jørgensen, "Mixing Oil and Water: Naturalizing Offshore Oil Platforms in Gulf Coast Aquariums," *Journal of American Studies* 46, no. 2 (2012): 461–80; Eva Hayward, "Sensational Jellyfish: Aquarium Affects and the Matter of Immersion," *differences* 23, no. 3 (2012): 161–96; Sam Muka, "Trashing the Tanks," *American Scientist* 106, no. 6 (2018): 340–44.

62 Jean Baudrillard, *Simulacra and Simulation* (Ann Arbor: University of Michigan Press, 1994).

63 Stephen Spotte, *Zoos in Postmodernism: Signs and Simulation* (Madison, NJ: Fairleigh Dickinson University Press, 2006).

64 Nyhart, *Modern Nature*; Christina Wessely and Nathan Stobaugh, "Watery Milieus: Marine Biology, Aquariums, and the Limits of Ecological Knowledge circa 1900," *Grey Room* 75 (2019): 36–59; Jørgensen, "Mixing Oil and Water."

65 Victor Hugo, *Toilers of the Sea*, trans. W. Moy Thomas and illustrated by Gustave Doré (London: Sampson Low, Son, and Marston 1867).

66 Ann Elias, *Coral Empire: Underwater Oceans, Colonial Tropics, Visual Modernity* (Durham, NC: Duke University Press, 2019).

67 Georgina Evans, "Framing Aquatic Life," *Screen* 61, no. 2 (2020): 170.

68 There are some cameras that do show the tank and the visitors, including Monterey's kelp forest and open sea cameras and all of the Mystic Aquarium beluga whale cameras.

69 Gregg Mitman, "Cinematic Nature: Hollywood Technology, Popular Culture, and the American Museum of Natural History," *Isis* 84, no. 4 (1993): 637–61; Jonathan Christopher Crylen, "The Cinematic Aquarium: A History of Undersea Film" (PhD diss., University of Iowa, 2015); Margaret Cohen, "The Underwater Imagination: From Environment to Film Set, 1954–1956," *English Language Notes* 57, no. 1 (2019): 51–71.

70 Mitman, "Cinematic Nature," 657. For a conversation on how filmmakers construct these underwater environments, see Eleanor Louson, "Never before Seen: Spectacle, Staging, and Story in Wildlife Film's Blue-Chip Renaissance. Dissertation" (PhD diss., York University [ON], 2018).

Chapter Two

1 Erwan Delrieu-Trottin, Jeffrey T. Williams, and Serge Planes, "*Macropharyngodon pakoko*, a New Species of Wrasse (Teleostei: Labridae) Endemic to the Marquesas Islands, French Polynesia," *Zootaxa* 3857, no. 3 (2014): 433–43. Interview with Jeff Williams, January 10, 2012.

2 Paul Lawrence Farber, *Finding Order in Nature: The Naturalist Tradition from Linnaeus to E.O. Wilson* (Baltimore: Johns Hopkins University Press, 2000).

3 Arthur MacGregor, *Naturalists in the Field: Collecting, Recording and Preserving the Natural World from the Fifteenth to the Twenty-First Century* (Leiden: Brill, 2018).

4 Warren D. Allmon, "The Evolution of Accuracy in Natural History Illustration: Reversal of Printed Illustrations of Snails and Crabs in Pre-Linnaean Works Suggests Indifference to Morphological Detail, *Archives of Natural History* 34, no. 1 (2007): 174–91. For a general history of scientific illustration, see Ann Shelby Blum, *Picturing Nature: American Nineteenth-Century Zoological Illustration* (Princeton, NJ: Princeton University Press, 1993), and David Knight *Zoological Illustration: An Essay Towards a History of Printed Zoological Pictures* (Folkstone, Kent: Wm Dawson and Son, 1977). For information on the early process of hand coloring images and wood printing, see Christine E.

Jackson, "The Materials and Methods of Hand-Colouring Zoological Illustrations," *Archives of Natural History* 38, no. 1 (2011): 53–64, and "The Painting of Hand-Coloured Zoological Illustrations," *Archives of Natural History* 38, no. 1 (2011): 36–52.

5 Samantha Muka, "Imagining the Sea: The Impact of Marine Field Work on Scientific Portraiture," in *Soundings and Crossings: Doing Science at Sea, 1800–1970*, ed. Katharine Anderson and Helen Rozwadowski (Sagamore Beach, MA: Science History, 2016), 247–76.

6 Isabel Cooper, "Artist at Large," *Atlantic Monthly*, July 1926, 91.

7 Isobel Armstrong, *Victorian Glassworlds: Glass Culture and the Imagination 1830–1880* (Oxford: Oxford University Press, 2008).

8 Interview with Jeff Williams, January 10, 2012.

9 Ibid.

10 Sue Dale Tunnicliffe and Annette Scheersoi. "Natural History Dioramas," in *Natural History Dioramas: History, Construction and Educational Role* (Dordrecht: Springer, 2015).

11 Adrienne Zihlman, "The Paleolithic Glass Ceiling: Women in Human Evolution," in *Women in Human Evolution*, ed. Lori D. Hager (London: Routledge, 1997).

12 Lynn K. Nyhart, *Modern Nature* (Chicago: University of Chicago Press, 2009); Ann Elias, *Coral Empire: Underwater Oceans, Colonial Tropics, Visual Modernity* (Durham, NC: Duke University Press, 2019).

13 Derek Bousé, *Wildlife Films* (Philadelphia: University of Pennsylvania Press, 2011). Paraphrased in Eleanor Louson, "Taking Spectacle Seriously: Wildlife Film and the Legacy of Natural History Display," *Science in Context* 31, no. 1 (2018): 16.

14 Kelley E. Wilder, "Photography and the Art of Science," *Visual Studies* 24, no. 2 (2009): 163–68; Jennifer Tucker, *Nature Exposed: Photography as Eyewitness in Victorian Science* (Baltimore: Johns Hopkins University Press, 2005).

15 Lorraine Daston and Peter Galison, *Objectivity* (Princeton, NJ: Princeton University Press, 2021).

16 Phillip Prodger, *Darwin's Camera: Art and Photography in the Theory of Evolution* (Oxford: Oxford University Press, 2009), xxiii.

17 Michael Lynch, "The Production of Scientific Images: Vision and Re-vision in the History, Philosophy, and Sociology of Science," in *Visual Cultures of Science: Rethinking Representational Practices in Knowledge Building and Science Communication*, ed. Luc Pauwels (Hanover, NH: Dartmouth College Press, 2006): 26–40; Nicolas Rasmussen, *Picture Control: The Electron Microscope and the Transformation of Biology in America, 1940–1960* (Stanford, CA: Stanford University Press, 1999).

18 Elias, *Coral Empire.*

19 "Filming the Impossible Sets: Filming Burrows and Tanks," BBC News, April 29, 2016, http://www.bbc.com/earth/story/20160310-filming-the-impossible-sets-filming-burrows-and-tanks (accessed December 12, 2019; no longer posted).

20 Alejandro Martínez, "'A Souvenir of Undersea Landscapes': Underwater Photography and the Limits of Photographic Visibility, 1890–1910." *História, Ciências, Saúde-Manguinhos* 21 (2014): 1029–47.

21 Edward Eigen, "On the Screen and in the Water: On Photographically Envisioning the Sea," *L'architecture, les sciences et la culture de l'histoire au XIXe siècle* (Saint-Etienne: Publications de l'Université de Saint-Etienne, 2001); Hanna Rose Shell, "Things under Water: Etienne-Jules Marey's Aquarium Laboratory and Cinema's Assembly," in Making Things Public: Atmospheres of Democracy, ed. Bruno Latour and Peter Weibel (Cambridge: MIT Press, 2005), 326–32; Jonathan Christopher Crylen, "The Cinematic Aquarium: A History of Undersea Film" (PhD diss., Iowa University, 2015).

22 Paul Louis Marie Fabre-Domergue and Eugène Biétrix, *Développement de la sole (*Solea vulgaris*): Introduction à l'étude de la pisciculture marine* (Paris: Vuibert and Nony, 1905).

23 Edward Eigen, "Dark Space and the Early Days of Photography as a Medium," *Grey Room* 3 (2001): 90–111.

24 P. Fabre-Domergue, *La photographie des animaux aquatiques* (Paris: George Carré et C. Naud, 1899), 1.

25 Ibid., 4.

26 Fabre-Domergue, *La photographie des animaux aquatiques.*

27 Ann Fabian, *The Skull Collectors: Race, Science, and America's Unburied Dead* (Chicago: University of Chicago Press, 2010); Samuel J. Redman, *Bone Rooms: From Scientific Racism to Human Prehistory in Museums* (Cambridge, MA: Harvard University Press, 2016).

28 R. W. Shufeldt, "Experiments in Photography of Live Fishes," *Bulletin of the United States Fish Commission, no. 424* (1899): 1–5, 9 plates.

29 Emily Pawley, "The Point of Perfection: Cattle Portraiture, Bloodlines, and the Meaning of Breeding, 1760–1860," *Journal of the Early Republic* 36, no. 1 (2016): 37–72; Harriet Ritvo, *The Animal Estate: The English and Other Creatures in the Victorian Age* (Cambridge, MA: Harvard University Press, 1987).

30 William Innes to William I. Homer, April 7, 1965, Folder 2, William Innes Collection, Academy of Natural Sciences of Drexel University.

31 Walter Lee Rosenberger to William T. Innes, April 24, 1917, Box 1, William T. Innes Papers, American Philosophical Society, Philadelphia.

32 Innes to Homer, April 7, 1965.

33 Ibid.

34 Innes, "Aquarium Fish Photography," *Complete Photographer* 1, no. 4 (1941): 238–48.

35 William Innes Papers, Academy of Natural Sciences of Drexel University, Philadelphia.

36 Myron Gordon to William T. Innes, February 17, 1930, William T. Innes, Box 1, "Myron Gordon" folder, American Philosophical Society Archives, Philadelphia.

37 George Sprague Myers to William T. Innes, August 12, 1931, Box 1, Folder A-Hom, William T. Innes Papers, American Philosophical Society Archives, Philadelphia,.

38 Sam C. Dunton, *Guide to Photographing Animals* (New York: Greenberg, 1956), 67.

39 Robert H. Boyle, "The Strange Fish and Stranger Times of Dr. Herbert R. Axelrod," *Sports Illustrated*, May 3, 1965. Axelrod's self-promotion and his legal troubles continued long after his spat with Innes. With the money he made from the growth and eventual sale of his pet publishing empire, Axelrod became a world-class violin collector. He eventually made a gift of many violins to the Smithsonian and the New Jersey Symphony Orchestra that was found to be overvalued at the time of gift for tax purposes. In addition, Axelrod was sentenced to eighteen years in federal prison in 2005 for tax fraud and evasion. See Ronald Smothers "Violin Collector Known for Sale to Orchestra Sentenced to 18 Months for Tax Fraud," *New York Times*, March 22, 2005.

40 Muka, "Imagining the Sea."

41 Lorus J. Milne, "A Simple, Thin Aquarium," *Science* 93, no. 2418 (1941): 432.

42 John E. Randall, "A Technique for Fish Photography," *Copeia*, no. 2 (1961): 241–42.

43 Roy Larson, "Obituary: Charles E. Cuttress, 1921–1992," *Bulletin of Marine Sciences* 51, no. 3 (1992): 480–81.

44 P. M. David, "The Photography of Life Oceanic Plankton Animals," *International Photo Tecknik* (1963): 40–43.

45 T. F. Pletcher, "A Portable Aquarium for Use at Sea to Photograph Fish and Aquatic Life," *Journal of the Fisheries Board of Canada* 23, no. 8 (1966): 1271–75.

46 Alan R. Emery and Richard Winterbottom, "A Technique for Fish Specimen Photography in the Field," *Canadian Journal of Zoology* 58, no. 11 (1980): 2162.

47 Steven Goodbred and Thomas Occhiogrosso, "Method for Photographing Small Fish," *Progressive Fish-Culturist* 41, no. 2 (1979): 76–77; Douglas G. McGrogan, "Mirror-Box for Photographing Small Fishes," *Copeia*, no. 4 (1990): 1174–76; Erling Holm, "Improved Technique for Fish Specimen Photography in the Field," *Canadian Journal of Zoology* 67, no. 9 (1989): 2329–

32; Jeffrey C. Howe, "A Technique for Immobilizing and Photographing Small, Live Fishes," *Fisheries Research* 27, no. 4 (1996): 261–64.

48 John N. Rinne and Martin D. Jakle, "The Photarium: A Device for Taking Natural Photographs of Live Fish," *Progressive Fish-Culturist* 43, no. 4 (1981): 201–4.

49 Ibid., 203.

50 Juergen Herler, Lovrenc Lipej, and Tihomir Makovec, "A Simple Technique for Digital Imaging of Live and Preserved Small Fish Specimens," *Cybium* 31, no. 1 (2007): 39–44; Dirk Steinke, Robert Hanner, and Paul D. N. Hebert, "Rapid High-Quality Imaging of Fishes Using a Flat-Bed Scanner," *Ichthyological Research* 56, no. 2 (2009): 210–11.

51 Alert Diver, "Expedition Twilight Zone," http://www.alertdiver.com/Expedition_Twilight_Zone (accessed February 15, 2017; no longer posted). See also Hollis Gear, https://www.facebook.com/hollisgear/posts/1187732164576750 (accessed March 8, 2022; no longer posted).

52 Ormestad, Mattias, Aldine Amiel, and Eric Röttinger, "Ex-situ Macro Photography of Marine Life," *Imaging Marine Life: Macrophotography and Microscopy Approaches for Marine Biology*, ed. Emmanuel Reynaud (Weinheim an der Bergstrasse: Wiley, 2013), 210–33.

53 Samantha Muka, "Taking Hobbyists Seriously," *Journal for the History and Philosophy of Biology* (forthcoming).

54 Elias, *Coral Empire*; Nyhart, *Modern Nature*.

55 D. W. Greenfield and J. E. Randall, "*Myersina balteata*, a New Shrimp-Associated Goby (Teleostei: Gobiidae) from Guadalcanal, Solomon Islands," *Journal of the Ocean Science Foundation* 30 (2018): 90–99.

Chapter Three

1 Alex Andon, "Jellyfish Tank by Jellyfish Art" Kickstarter.com. https://www.kickstarter.com/projects/jellyfishart/jellyfish%ADaquarium/description.

2 Observer Staff, "Jellyfish Tanks, Funded 54 Times over on Kickstarter, Turn Out to Be Jellyfish Death Trap," Observer.com, March 15, 2012, http://observer.com/2012/03/jellyfishtanksfunded54timesoveronkickstarter turnouttobejellyfishdeathtraps(no longer posted); Alex Andon, "Desktop Jellyfish Tank-Alex," Kickstarter.com. https://www.kickstarter.com/projects/1497255984/desktop-jellyfish-tank/.

3 "Introduction: A Guide to Caring for Your Moon Jellyfish," JellyfishCare.com, http://www.jellyfishcare.com/; "Moon Jellyfish Blog: Tips and Tricks to Keeping Jellyfish as Pets," Sunset Marine Labs, https://moonjellyfishblog.com.

4 JellyfishArtDotCom, "Jellyfish Aquarium Kickstarter w Vanilla Ice," 2017, https://www.youtube.com/watch?v=7LbC7_ZhrJQ; "Jellyfishart/EON/

Cubic Tank Discussion and Fan Page," https://www.facebook.com/groups/397225970290084/search/?query=problem.

5 Monterey Bay Aquarium, "Comb Jelly," Animals A to Z, https://www.montereybayaquarium.org/animals-and-exhibits/animal-guide/invertebrates/comb-jelly.

6 The term *jellyfish* is not scientific, but it is useful. *Jellyfish* or *jelly* is a term used to refer to gelatinous organisms that resemble adult cnidaria, with a recognizable bell shape and pulsing movement. Although there are several different families of organisms that get lumped under the term *jelly*, it is a useful term here because most of these organisms require similar living conditions regardless of their taxonomic grouping. In places where the species or specificity matter, I have used the Latin name or made clarifications. But I will use the term *jellyfish* throughout the chapter as a shorthand.

7 Samantha Muka, "'A New York Institution': The Impact of the New York Aquarium on Biological Research in the New York Area, 1898–1967" (forthcoming).

8 Frank Nutch, "Gadgets, Gizmos, and Instruments: Science for the Tinkering," *Science, Technology, & Human Values* 21, no. 2 (1996): 214–28; Lisa M. Frehill, "The Gendered Construction of the Engineering Profession in the United States, 1893–1920," *Men and Masculinities* 6, no. 4 (2004): 383–403.

9 Clinton R. Sanders, "Working Out Back: The Veterinary Technician and 'Dirty Work,'" *Journal of Contemporary Ethnography* 39, no. 3 (2010): 243–72; Hilary Rose, "Hand, Brain, and Heart: A Feminist Epistemology for the Natural Sciences," *Signs* 9, no. 1 (1983): 73–90.

10 Joan W. Scott, "Gender: A Useful Category of Historical Analysis," *American Historical Review* 91, no. 5 (1986): 1053–75; Gayle Rubin, "The Traffic in Women: Notes on the 'Political Economy' of Sex," *Toward an Anthropology of Women*, ed. Rayna Rapp (New York: Monthly Review Press, 1975), 157–210.

11 This simplified life cycle does not necessarily need to be lived cyclically. There are some species of jelly, including the aptly named "immortal jellyfish," that can go backwards in this cycle. After it becomes a medusa, if it is injured, it can revert to polyp form to reset its life cycle. L. Martell, S. Piraino, C. Gravili, and F. Boero, "Life Cycle, Morphology and Medusa Ontogenesis of *Turritopsis dohrnii* (Cnidaria: Hydrozoa)," *Italian Journal of Zoology* 83, no. 3 (2016): 390–99.

12 Lisa-Ann Gershwin, *Stung! On Jellyfish Blooms and the Future of the Ocean* (Chicago: University of Chicago Press, 2013).

13 Samantha K. Muka, "The Right Tool and the Right Place for the Job: The Importance of the Field in Experimental Neurophysiology, 1880–1945," *History and Philosophy of the Life Sciences* 38, no. 3 (2016): 1–28; "Illuminating Animal Behavior: The Impact of Malleable Marine Stations on Tropism Research," in *From the Beach to the Bench: Why Marine Biological Studies?*, ed.

Jane Maienschein, Karl Matlin, and Rachel Ankeny (Chicago: University of Chicago Press, 2020), 119–43.

14 Edward T. Browne, "On Keeping Medusae Alive in an Aquarium," *Journal of the Marine Biological Association of the United Kingdom* 5, no. 2 (1898): 176.

15 S. Kemp and A. V. Hill, "Edgar Johnson Allen. 1866–1942," *Obituary Notices of Fellows of the Royal Society* 4, no. 12 (1943): 361.

16 Douglas P. Wilson, "The Plunger-Jar," *Aquarist and Pondkeeper* 1 (1937): 138–39.

17 Browne, "Keeping Medusae Alive," 179.

18 Ibid., 178; F. G. Walton Smith, "An Apparatus for Rearing Marine Organisms in the Laboratory" *Nature* (August 31, 1935): 345–46. Eventually, experimentalists found that jellyfish required constantly circulating water, not only because of muscle exhaustion but also because they produce copious amounts of mucous when they come into contact with other organisms, especially when they feed.

19 Samantha K. Muka, "The Right Tool and the Right Place for the Job: The Importance of the Field in Experimental Neurophysiology, 1880–1945," *History and Philosophy of the Life Sciences* 38, no. 3 (2016): 1–28.; Muka, "Illuminating Animal Behavior."

20 W. J. Rees and F. S. Russell, "On Rearing the Hydroids of Certain Medusae, with an Account of the Methods Used," *Journal of the Marine Biological Association of the United Kingdom* 22, no. 1 (1937): 61–82.

21 Maude J. Delap, "Notes on the Rearing of *Chrysaora isosceles* in an Aquarium," *Irish Naturalist* 10, no. 2 (February 1901): 25–28; N. F. McMillan and W. J. Rees, "Maude Jane Delap," *Irish Naturalists' Journal* 12, no. 9 (January 1958): 221–22.

22 Delap, "Notes on the Rearing of *Chrysaora isosceles*," 25.

23 Ibid., 27.

24 Ibid.

25 Maude J. Delap, "Notes on the Rearing, in an Aquarium, of *Cyanea lamarcki*, Péron et Lesueur," *Report on the Sea and Inland Fisheries of Ireland for 1905* (Dublin: n.p., 1907), 20–22; Maude J. Delap. "Notes on the Rearing, in an Aquarium, of *Aurelia aurita*, L. and *Pelagia perla* (Slabber)," *Report on the Sea and Inland Fisheries of Ireland for 1905* (Dublin: n.p., 1907), 160–64, and 2 plates.

26 Marie V. Lebour, "The Food of Plankton Organisms. II," *Journal of the Marine Biological Association of the United Kingdom* 13, no. 1 (1923), 75. See also Marie V. Lebour, "The Food of Plankton Organisms," *Journal of the Marine Biological Association of the United Kingdom* 12 (1922): 644–77; F. S. Russell, "Dr. Marie V. Lebour," *Journal of the Marine Biological Association of the United Kingdom* 52, no. 3 (August 1972): 777–88.

27 F. G. Gilchrist "Rearing the Scyphistoma of *Aurelia* in the Laboratory" in *Culture Methods for Invertebrate Animals: A compendium prepared cooperatively by American zoologists under the direction of a committee from Section F of the American Association for the Advancement of Science*, ed. Frank E. Lutz, Paul L. Welch and Paul S. Galtsoff (New York: Dover, 1937), 143.

28 Wulf Greve, "The 'Planktonkreisel': A New Device for Culturing Zooplankton," *Marine Biology* 1, no. 3 (1968): 201–3. A note about capitalization: In German, the name of the tank would be capitalized as Planktonkreisel. However, Greve does not always capitalize the tank names and usually only does so at the beginning of his published writings. The marine network community does not capitalize the names of these tanks, and therefore I have chosen to leave them lowercase after the first mention of them in the text.

29 Wulf Greve, "Cultivation Experiments on North Sea Ctenophores," *Helgoländer wissenschaftliche Meeresuntersuchungen* 20, no. 1 (1970): 304.

30 Wulf Greve, "The 'Meteor Planktonküvette': A Device for the Maintenance of Macrozooplankton Aboard Ships," *Aquaculture* 6, no. 1 (1975): 77–82.

31 Greve, "Cultivation Experiments," 310.

32 M. M. Kamshilov, "The Dependence of Ctenophore *Beroe cucumis* Fab Sizes from Feeding," *Doklady of the Academy of Sciences of the USSR* 131 (1960): 957–60. 1960; "Pitanie Grebnevika Beroe-Cucumis Fab," *Doklady of the Academy of Sciences of the USSR* 102, no. 2 (1955): 399–402.

33 N. Swanberg, "The Feeding Behavior of *Beroe ovata*," *Marine Biology* 24, no. 1 (1974): 69–76.

34 L. D. Baker and M. R. Reeve, "Laboratory Culture of the Lobate Ctenophore *Mnemiopsis mccradyi* with Notes on Feeding and Fecundity," *Marine Biology* 26, no. 1 (1974): 57–62.

35 A. K. Nagabhushanam, "Feeding of a Ctenophore, *Bolinopsis infundibulum* (O. F. Müller)," *Nature* 184, no. 4689 (1959): 829–29.

36 Jed Hirota, "Laboratory Culture and Metabolism of the Planktonic Ctenophore, *Pleurobrachia bachei* A. Agassiz," in *Biological Oceanography of the Northern North Pacific Ocean*, ed. A. Y. Takenouti (Tokyo: Idemitsu Shoten, 1972).

37 Hirokazu Matsuda and Taisuke Takenouchi, "Development of Technology for Larval Culture in Japan: A Review," *Bulletin of the Fisheries Research Agency* 20 (2007): 77–84; Fisheries Biologists still work with kreisel tanks for lobster culture. For information, see Jason S. Goldstein and Brian Nelson, "Application of a Gelatinous Zooplankton Tank for the Mass Production of Larval Caribbean Spiny Lobster, *Panulirus argus*," *Aquatic Living Resources* 24, no. 1 (2011): 45–51.

38 Etur Hirai, "On the Developmental Cycles of *Aurelia aurita* and *Doctylometra pacifica*," *Bulletin of the Marine Biological Station of Asamushi, Tohoku*

University 9 (1958): 81; "On the Species of *Cladonema radiatum* var. *mayeri* Perkins" *Bulletin of the Marine Biological Station of Asamushi, Tohoku University* 9 (1958): 23–25. Both of these papers are published under the single authorship of Hirai, but he acknowledges Kakinuma as a graduate student who does equal work in the paper.

39 Y. Kakinuma, "An Experimental Study of the Life Cycle and Organ Differentiation of *Aurelia aurita* Lamarck," *Bulletin of the Marine Biological Station of Asamushi, Tohoku University* 15 (1975): 101–12.

40 Abe Yoshihisa and M. Hisada, "On a New Rearing Method of Common Jellyfish, *Aurelia aurita*," *Bulletin of the Marine Biological Station of Asamushi, Tohoku University* 13, no. 3 (1969): 205–9.

41 Dorothy Breslin Spangenberg, "A Study of Strobilation in *Aurelia aurita* under Controlled Conditions," *Journal of Experimental Zoology Part A: Ecological Genetics and Physiology* 160, no. 1 (1965): 1–9; "Cultivation of the Life Stages of *Aurelia aurita* under Controlled Conditions," *Journal of Experimental Zoology Part A: Ecological Genetics and Physiology* 159, no. 3 (1965): 303–18; "Iodine Induction of Metamorphosis in *Aurelia*," *Journal of Experimental Zoology Part A: Ecological Genetics and Physiology* 165, no. 3 (1967): 441–49.

42 Muka, "Right Tool."

43 Jürgen Lange and Rainer Kaiser, "The Maintenance of Pelagic Jellyfish in the Zoo-Aquarium Berlin," *International Zoo Yearbook* 34, no. 1 (1995): 60.

44 William M. Hamner, "Design Developments in the Planktonkreisel, a Plankton Aquarium for Ships at Sea," *Journal of Plankton Research* 12, no. 2 (1990): 397.

45 F. A. Sommer, "Advances in Culture and Display of *Aurelia aurita*, the Moon Jelly." In *AAZPA Regional Conference Proceedings* (N.p.: American Association of Zoological Parks and Aquariums, 1992), 391–96.

46 David C. Powell, *A Fascination for Fish: Adventures of an Underwater Pioneer* (Berkeley: University of California Press, 2001).

47 Ibid.

48 Ibid., 162.

49 Lisa-Ann Gershwin and Allen G. Collins, "A Preliminary Phylogeny of Pelagiidae (Cnidaria, Scyphozoa), with New Observations of *Chrysaora colorata* comb. nov.," *Journal of Natural History* 36, no. 2 (2002): 127–48.

50 F. A. Sommer, "Jellyfish and Beyond: Husbandry of Gelatinous Zooplankton at the Monterey Bay Aquarium," in *Proceedings of the Third International Aquarium Congress*, ed. Chris Barrett (Boston: New England Aquarium, 1993), 249–61; "Husbandry Aspects of a Jellyfish Exhibit at the Monterey Bay Aquarium," *American Association of Zoological Parks and Aquariums Annual Conference Proceedings* (1992): 362–69.

51 Sommer, "Husbandry Aspects" 365; "Jellyfish and Beyond," 256–58.

52 One of Shimura's few publications is Kazuko Shimura, Tanimura Shunsuke, and Shimazu Tsuneo, "Breeding of *Dactylometra pacifica* in *Enoshima aquarium*," *Journal of Japanese Zoos and Aquariums* 30, no. 3 (1988), 76–79.

53 Powell, *A Fascination for Fish*, 162.

54 Rebecca R. Helm, "New Research Reveals How to Easily Grow Jellyfish in Captivity," *Deep Sea News*, December 28, 2017, https://www.deepseanews.com/2017/12/new-research-reveals-how-to-easily-grow-jellyfish-in-captivity/.

55 Rebecca R. Helm and Casey W. Dunn, "Indoles Induce Metamorphosis in a Broad Diversity of Jellyfish, but Not in a Crown Jelly (Coronatae)," *PloS One* 12, no. 12 (2017): e0188601.

56 Sommer, "Husbandry Aspects," 362.

57 Monterey Bay Aquarium, "Behind the Scenes at Jelly School," https://montereybayaquarium.tumblr.com/post/143489586643/behind-the-scenes-at-jelly-school.

58 Chad L. Widmer, *How to Keep Jellyfish in Aquariums: An Introductory Guide for Maintaining Healthy Jellies* (Tucson, AZ: Wheatmark, 2008), xiv–xv.

59 Ibid., xiv.

Chapter Four

1 This work was made possible by a postdoctoral fellowship from the Smithsonian Institution and the mentorship of Pamela Henson (Smithsonian Institutional Archives) and Nancy Knowlton (National Museum of Natural History).

2 Samantha Muka, "The Evolution of a Reef Aquarium." Smithsonian Institution, Ocean, October 2017, https://ocean.si.edu/ecosystems/coral-reefs/evolution-reef-aquarium. Adey used the terms *mesocosm* and *microcosm* to describe his particular systems. *Microcosm* describes a simplified experimental ecosystem used in the laboratory. Because it is a term not generally used outside of academic biology, I do not describe all tanks as *microcosms* and only use the word when the tank builder uses it. Walter H. Adey, "The Microcosm: A New Tool for Reef Research," *Coral Reefs* 1, no. 3 (1983): 193–201.

3 Walter Adey and Karen Loveland, *Dynamic Aquaria: Building Living Ecosystems* (San Diego, CA: Academic Press, 1998), 1.

4 Julian Sprung, *Reef Notes 3: 1993/1994*, revisited and revised ed. (Coconut Grove, FL: Ricordea, 1996).

5 Humblefish, "A Tribute to Lee Chin Eng!," Nano-Reef.com, April 21, 2020, https://www.nano-reef.com/forums/topic/412435-a-tribute-to-lee-chin-eng/.

6 David R. Hershey, "Doctor Ward's Accidental Terrarium," *American Biology Teacher* 58, no. 5 (1996): 276–81; Vernon N. Kisling, ed., *Zoo and Aquarium*

History: Ancient Animal Collections to Zoological Gardens (Boca Raton, FL: CRC, 2000).

7 Lee Chin Eng, “Nature’s System of Keeping Marine Fishes,” *Tropical Fish Hobbyist* 9, no. 6 (1961): 23–30.

8 Adey and Loveland, *Dynamic Aquaria*.

9 “Live Rock Hitchhikers,” ARC Reef Marine Research Laboratory, March 2, 2020, https://arcreef.com/live-rock-hitchhikers/. The ARC website lists common hitchhikers on Atlantic rock purchased from ARC Reef directly. There are more comprehensive live rock hitchhiker lists on other websites, including Crabs McJones, “What Is That!! A R2R Guide to Common New Tank Hitchhikers,” Reef2Reef, Hitchhiker and Critter ID, https://www.reef2reef.com/threads/what-is-that-a-r2r-guide-to-common-new-tank-hitchhikers.443382/.

10 Patrick Schubert and Thomas Wilke, “Coral Microcosms: Challenges and Opportunities for Global Change Biology,” in *Corals in a Changing World*, ed. Carmenza Duque and Edisson Tello Camacho (N.p.: IntechOpen, 2017), 143.

11 Greta S. Aeby, Blake Ushijima, Justin E. Campbell, Scott Jones, Gareth J. Williams, Julie L. Meyer, Claudia Häse, and Valerie J. Paul, “Pathogenesis of a Tissue Loss Disease Affecting Multiple Species of Corals along the Florida Reef Tract,” *Frontiers in Marine Science* 6 (2019): 678; Ushijima, Blake, Julie L. Meyer, Sharon Thompson, Kelly Pitts, Michael F. Marusich, Jessica Tittl, Elizabeth Weatherup, et al., “Disease Diagnostics and Potential Coinfections by *Vibrio coralliilyticus* during an Ongoing Coral Disease Outbreak in Florida,” *Frontiers in Microbiology* October 26, 2020, https://www.frontiersin.org/articles/10.3389/fmicb.2020.569354/full.

12 Bartlett, Thomas C. “Small Scale Experimental Systems for Coral Research: Considerations, Planning, and Recommendations,” in *NOAA Technical Memorandum NOS NCCOS* (2013).

13 Alistair Sponsel, *Darwin’s Evolving Identity: Adventure, Ambition, and the Sin of Speculation* (Chicago: University of Chicago Press, 2018); David R. Stoddart, “Darwin, Lyell, and the Geological Significance of Coral Reefs,” *British Journal for the History of Science* 9, no. 2 (1976): 199–218; James Bowen, *The Coral Reef Era: From Discovery to Decline; A History of Scientific Investigation from 1600 to the Anthropocene Epoch* (New York: Springer, 2015); Brian Roy Rosen, “Darwin, Coral Reefs, and Global Geology,” *BioScience* 32, no. 6 (1982): 519–25.

14 Patrick L. Colin, “A Brief History of the Tortugas Marine Laboratory and the Department of Marine Biology, Carnegie Institution of Washington,” in *Oceanography: The Past*, ed. M Sears and D. Merriman (New York: Springer, 1980), 138–47; Elizabeth N. Shor, “The Role of T. Wayland Vaughan in American Oceanography,” in *Oceanography: The Past*, ed. M Sears and D. Merriman (New York: Springer, 1980), 127–37; Lester D. Stephens and

Dale R. Calder, *Seafaring Scientist: Alfred Goldsborough Mayor, Pioneer in Marine Biology* (Columbia: University of South Carolina Press, 2006).

15 Vaughan, Thomas Wayland. "The Results of Investigations of the Ecology of the Floridian and Bahaman Shoal-Water Corals," *Proceedings of the National Academy of Sciences* 2, no. 2 (1916): 97.

16 "On the Rate of Growth of Stony Corals," Folders 1–3, Box 6, Accession 99–124, T. Wayland Vaughan Papers, Smithsonian Institution Archives, Washington, DC. See also Thomas Wayland Vaughan, "The Geologic Significance of the Growth-Rate of the Floridian and Bahaman Shoal-Water Corals," *Journal of the Washington Academy of Sciences* 5, no. 17 (1915): 591–600.

17 Charles Maurice Yonge, *A Year on the Great Barrier Reef: The Story of Corals and of the Greatest of Their Creations* (New York: Putnam, 1930); *Great Barrier Reef Expedition, 1928–1929* (London: British Museum, 1930).

18 Marion Endt-Jones, "'Something Rich and Strange': Coral in Contemporary Art," *Framing the Ocean, 1700 to the Present: Envisaging the Sea as Social Space*, ed. Tricia Cusack (London: Routledge, 2017), 223–238; Katharine Anderson, "Coral Jewelry," *Victorian Review* 34, no. 1 (2008): 47–52; Matt K. Matsuda, *Pacific Worlds: A History of Seas, Peoples, and Cultures* (Cambridge: Cambridge University Press, 2012); Pedro Machado, Steve Mullins, and Joseph Christensen, eds. *Pearls, People, and Power: Pearling and Indian Ocean Worlds* (Athens: Ohio University Press, 2020).

19 Ann Elias, *Coral Empire: Underwater Oceans, Colonial Tropics, Visual Modernity* (Durham, NC: Duke University Press, 2019), 71.

20 Howard T. Odum and Eugene P. Odum, "Trophic Structure and Productivity of a Windward Coral Reef Community on Eniwetok Atoll," *Ecological monographs* 25, no. 3 (1955): 291.

21 Ibid. 310.

22 I refer to René and Ida as René and Ida Catala-Stucki. While René published under his own last name, he insisted that his wife hyphenate their names after marriage. Ida was an integral partner in the development of tank craft, and she worked closely with René to found the Noumea aquarium. However, she is rarely mentioned in his work and usually only as "my wife." Ida does not often appear in literature about the aquarium, but in a travel story by Anaïs Nin she is noted as a Dr. Catala Stucki, an "oceanographer, a scientist, and a deep-sea diver." Anaïs Nin, *In Favor of the Sensitive Man and Other Essays* (San Diego: Harcourt Brace, 1966), 162.

23 René Catala, *Carnival under the Sea*(Paris: R. Sicard, 1964).

24 Rozwadowski, Helen. "Playing By—and on and under—the Sea: The Importance of Play for Knowing the Ocean," in *Knowing Global Environments: New Historical Perspectives on the Field Sciences*, ed. Jeremy Vetter (New Brunswick, NJ: Rutgers University Press, 2010), 162–89.

25 Catala, *Carnival under the Sea,* 10.

26 Ibid., 19.

27 Lee Chin Eng, "Nature's System of Keeping Marine Fishes," *Tropical Fish Hobbyist* 9, no. 6 (1961): 26.

28 Ibid.

29 Ibid. 25.

30 Paul L. Jokiel, "Solar Ultraviolet Radiation and Coral Reef Epifauna," *Science* 207, no. 4435 (1980): 1069–71

31 S. L. Coles and Paul L. Jokiel, "Synergistic Effects of Temperature, Salinity and Light on the Hermatypic Coral *Montipora verrucosa*," *Marine Biology* 49, no. 3 (1978): 187–95.

32 Stephen L. Coles, Paul L. Jokiel, and Clark R. Lewis, "Thermal Tolerance in Tropical versus Subtropical Pacific Reef Corals," *Pacific Science* 30, no. 2 (1976): 159–66.

33 Bruce A. Carlson, "General Introduction: Advances in Coral Husbandry in Public Aquaria," in *Advances in Coral Husbandry in Public Aquariums*, Public Aquarium Husbandry Series, vol. 2, ed. R. J. Leewis and M. Janse (Arnhem, The Netherlands: Burgers' Zoo, 2008), ix–xv.

34 *Living rock* is the term most commonly used by aquarists today to refer to rock that has not been treated to remove bacteria and inhabitants before placing it into the aquarium. Wilkens is the first to use the term *living rock*, but both the Catala-Stuckis and Eng suggested using untreated rock in their systems.

35 Bruce A. Carlson, "Aquarium Systems for Living Cora," *International Zoo Yearbook* 26, no. 1 (1987): 3

36 Peter Wilkens, *The Saltwater Aquarium for Tropical Marine Invertebrates* (Berlin: Engelbert Pfriem, 1973), 77.

37 Ibid. 73.

38 Peter Wilkens, "Mini-Reef," *Marine Aquarist* 7, no. 5 (1976): 37.

39 Ibid., 39.

40 Ibid., 42.

41 Ibid.

42 "The Father of Modern Reef Keeping: Lee Chin Eng," *Reef Aquarium Farming News*, no. 6 (June 1997): 3, http://www.garf.org/news6p3.html.

43 Jake Adams, "Historic *Stuber acropora* Photograph Documents 25 Years of Stony Coral Reefing," Reef Builders, News, January 5, 2011, https://reefbuilders.com/2011/01/05/historic-stuber-acropora-photograph-documents-25-years-stony-coral-reefing/.

44 Oceans, Reefs, and Aquariums (ORA), "Grube's Gorgonian—New from ORA," October 30, 2011, https://www.orafarm.com/blog/2011/10/30/grubes-gorgonian-new-from-ora/.

45 Peter Wilkens, "An Experimental Marine Aquarium," *Marine Aquarist* 6, no. 5 (1975): 49–55.

46 J. Jaubert, "An Integrated Nitrifying-Denitrifying Biological System Capable of Purifying Sea Water in a Closed Circuit Aquarium," *Bulletin de l'Institut. Océanographique, Monaco* 5 (1989): 101–6.

47 Ibid., 105.

48 Jean M. Jaubert, "Scientific Considerations on a Technique of Ecological Purification That Made Possible the Cultivation of Reef Building Corals in Monaco," in *Advances in Coral Husbandry in Public Aquariums*, ed. Rob J. Leewis and Max Janse (Arnhem, The Netherlands: Burgers' Zoo, 2008), 116.

49 Nicolas Leclercq, Jean-Pierre Gattuso, and Jean Jaubert, "Primary Production, Respiration, and Calcification of a Coral Reef Mesocosm under Increased CO_2 Partial Pressure," *Limnology and Oceanography* 47, no. 2 (2002): 558–64; Stéphanie Reynaud, Nicolas Leclercq, Samantha Romaine-Lioud, Christine Ferrier-Pagés, Jean Jaubert, and Jean-Pierre Gattuso, "Interacting Effects of CO_2 Partial Pressure and Temperature on Photosynthesis and Calcification in a Scleractinian Coral," *Global Change Biology* 9, no. 11 (2003): 1660–68.

50 Jean M. Jaubert, "Scientific Considerations on a Technique of Ecological Purification That Made Possible the Cultivation of Reef-Building Corals in Monaco," *Advances in Coral Husbandry in Public Aquariums*, ed. Rob J. Leewis and Max Janse (Arnhem, The Netherlands: Burgers' Zoo, 2008), 116.

51 Paul Jokiel does not appear in many citations detailing the history of tank building and is not cited by Wilkens, Adey, or Jaubert. It is possible that his work, which was published primarily in academic journals, went unnoticed by the larger hobbyist and aquarist community.

52 Box 10, Coral Reefs, Roll #135, Marine System Lab, June 7, 1982, 16 mm, Smithsonian Institution Archives, Accession 05–263, Smithsonian Productions.

53 Adey and Loveland, *Dynamic Aquaria*, 73.

54 Adey and Loveland, *Dynamic Aquaria*, 71.

55 Michael F. Canino, Ingrid B. Spies, Kathryn M. Cunningham, Lorenz Hauser, and W. Stewart Grant, "Multiple Ice-Age Refugia in Pacific Cod, *Gadus macrocephalus*," *Molecular Ecology* 19, no. 19 (2010): 4339–51.

56 Adey and Loveland, *Dynamic Aquaria*, 187.

57 Sprung, *Reef Notes* 3, 9.

58 Interview with Bill Hoffman, St. Lucie County Aquarium, Fort Pierce, Florida, August 9, 2017.

59 Christopher Luckett, Walter H. Adey, Janice Morrissey, and Donald M. Spoon, "Coral Reef Mesocosms and Microcosms: Successes, Problems, and the Future of Laboratory Models," *Ecological Engineering* 6, nos. 1–3 (1996): 60.

60 Jaubert, "Scientific Considerations on a Technique of Ecological Purification," 119.

61 Richard Ross, "Everyone Can Do Science," Skeptical Reefkeeping, *Reefs Magazine*, June 2, 2016, http://packedhead.net/2016/skeptical-reefkeeping-xiv-everyone-can-do-science/

62 Bruce Carlson, "Organism Responses to Rapid Change: What Aquaria Tell Us About Nature," *American Zoologist* 39 no. 1 (1999): 53.

63 Jelle Bijma, Hans-O. Pörtner, Chris Yesson, and Alex D. Rogers, "Climate Change and the Oceans–What Does the Future Hold?," *Marine Pollution Bulletin* 74, no. 2 (2013): 495–505; Tim P. Barnett, David W. Pierce, and Reiner Schnur, "Detection of Anthropogenic Climate Change in the World's Oceans," *Science* 292, no. 5515 (2001): 270–74; Jeffrey Maynard, Ruben Van Hooidonk, C. Mark Eakin, Marjetta Puotinen, Melissa Garren, Gareth Williams, Scott F. Heron, et al., "Projections of Climate Conditions That Increase Coral Disease Susceptibility and Pathogen Abundance and Virulence," *Nature Climate Change* 5, no. 7 (2015): 688–94; John F. Bruno, Elizabeth R. Selig, Kenneth S. Casey, Cathie A. Page, Bette L. Willis, C. Drew Harvell, Hugh Sweatman, and Amy M. Melendy, "Thermal Stress and Coral Cover as Drivers of Coral Disease Outbreaks," *PLoS Biology* 5, no. 6 (2007): e124; Ove Hoegh-Guldberg, Elvira S. Poloczanska, William Skirving, and Sophie Dove, "Coral Reef Ecosystems under Climate Change and Ocean Acidification," *Frontiers in Marine Science* 4 (2017): 158; Ellycia R. Harrould-Kolieb and Dorothée Herr, "Ocean Acidification and Climate Change: Synergies and Challenges of Addressing Both under the UNFCCC," *Climate Policy* 12, no. 3 (2012): 378–89.

64 M. Nonaka, A. H. Baird, T. Kamiki, and H. H. Yamamoto. "Reseeding the Reefs of Okinawa with the Larvae of Captive-Bred Corals," *Coral Reefs* 22, no. 1 (2003): 34.

65 Kristen L. Marhaver, Mark J. A. Vermeij, and Mónica M. Medina, "Reproductive Natural History and Successful Juvenile Propagation of the Threatened Caribbean Pillar Coral *Dendrogyra cylindrus*," *BMC Ecology* 15, no. 1 (2015): 9.

66 Florida Aquarium, "The Florida Aquarium Becomes the First Organization in History to Induce Spawning of Atlantic Coral," https://www.flaquarium.org/pressroom/posts/the-florida-aquarium-becomes-first-organization-in-history-to-induce-spawning-of-atlantic-coral-a-ne accessed; Karen Lynn Neely and Cynthia Lewis, "Rapid Population Decline of the Pillar Coral *Dendrogyra cylindrus* along the Florida Reef Tract," *bioRxiv* 5, no. 10 (2020): 434; Karen L. Neely, Cynthia Lewis, A. N. Chan, and I. B. Baums, "Hermaphroditic Spawning by the Gonochoric Pillar Coral *Dendrogyra cylindrus*," *Coral Reefs* 37, no. 4 (2018): 1087–92.

67 Netta Kooperman, Eitan Ben-Dov, Esti Kramarsky-Winter, Zeev Barak, and Ariel Kushmaro, "Coral Mucus-Associated Bacterial Communities from

Natural and Aquarium Environments," *FEMS Microbiology Letters* 276, no. 1 (2007): 106–13; S. A. Gignoux-Wolfsohn, Christopher J. Marks, and Steven V. Vollmer, "White Band Disease Transmission in the Threatened Coral, *Acropora cervicor nis*," *Scientific reports* 2 (2012): 1–3; Ilsa B. Kuffner, Linda J. Walters, Mikel A. Becerro, Valerie J. Paul, Raphael Ritson-Williams, and Kevin S. Beach, "Inhibition of Coral Recruitment by Macroalgae and Cyanobacteria," *Marine Ecology Progress Series* 323 (2006): 107–17.

68 Madeleine J. H. van Oppen, James K. Oliver, Hollie M. Putnam, and Ruth D. Gates, "Building Coral Reef Resilience through Assisted Evolution," *Proceedings of the National Academy of Sciences* 112, no. 8 (2015): 2307–13.

69 Ben A. Minteer, Jane Maienschein, and J. P. Collins, eds., *The Ark and Beyond: The Evolution of Aquarium and Zoo Conservation* (Chicago: University of Chicago Press, 2018).

70 Emma Hayman and Peter Hannam, "'Coral Ark' Bids to Save Biodiversity as Global Threats to Reefs Mount," *Sydney Morning Herald*, December 15, 2019, https://www.smh.com.au/environment/sustainability/coral-ark-bids-to-save-biodiversity-as-global-threats-to-reefs-mount-20191213-p53jw8.html?fbclid=IwAR1xavP8O4jOw3dm22lWu3g9JKow-4gOH8nXcrNfN_Ad0evuWF4ai8_Ox0k.

71 To revisit the Beebe story, see chapter 1.

Chapter Five

1 Elizabeth Hanson, *Animal Attractions: Nature on Display in American Zoos* (Princeton, NJ: University Press, 2004); Lynn K. Nyhart, *Modern Nature* (Chicago: University of Chicago Press, 2009).

2 Samantha K. Muka, "Conservation Constellations: Aquariums in Aquatic Conservation Networks," in *The Ark and Beyond: The Evolution of Aquarium and Zoo Conservation*, ed. by Ben A. Minteer, Jane Maienschein, and James P. Collins (Chicago: University of Chicago Press, 2018), 90–104.

3 Ditch Townsend, "Sustainability, Equity and Welfare: A Review of the Tropical Marine Ornamental Fish Trade," *SPC Live Reef Fish Information Bulletin* 20 (December 2011): 2.

4 Gregg Yan, "Saving Nemo–Reducing Mortality Rates of Wild-Caught Ornamental Fish," *SPC Live Reef Fish Information Bulletin* 21 (2016): 3–7; Elizabeth Wood, "Collection of Coral Reef Fish for Aquaria: Global Trade, Conservation Issues and Management Strategies" Marine Conservation Society, UK, January 2001.

5 Martin A. Moe, *Breeding the Orchid Dottyback,* Pseudochromis fridmani: *An Aquarist's Journal*" (Plantation, FL: Green Turtle, 1997) 119.

6 Heather J. Koldewey and Keith M. Martin-Smith, "A Global Review of Seahorse Aquaculture," *Aquaculture* 302, no. 3/4 (2010): 131–52.

7 Simon Pouil, Michael F. Tlusty, Andrew L. Rhyne, and Marc Metian, "Aquaculture of Marine Ornamental Fish: Overview of the Production Trends and the Role of Academia in Research Progress," *Reviews in Aquaculture* 12, no. 2 (2020): 1217–30.

8 Robert E. Kohler, *Lords of the Fly: Drosophila Genetics and the Experimental Life* (Chicago: University of Chicago Press, 1994); Adele E. Clarke and Joan H. Fujimura, eds., *The Right Tools for the Job: At Work in Twentieth-Century Life Sciences* (Princeton, NJ: Princeton University Press, 2014); Karen Rader, *Making Mice: Standardizing Animals for American Biomedical Research, 1900–1955* (Princeton, NJ: Princeton University Press, 2004); Donna J. Haraway, *Modest_Witness@ Second_Millennium: FemaleMan_Meets_OncoMouse: Feminism and Technoscience*, with a preface and study guide by Thyrza Nichols Goodeve (New York: Routledge, 2018).

9 Kathy J. Cooke, "From Science to Practice, or Practice to Science? Chickens and Eggs in Raymond Pearl's Agricultural Breeding Research, 1907–1916," *Isis* 88, no. 1 (1997): 62–86; William Boyd, "Making Meat: Science, Technology, and American Poultry Production," *Technology and Culture* 42, no. 4 (2001): 631–64; Emily Pawley, "The Point of Perfection: Cattle Portraiture, Bloodlines, and the Meaning of Breeding, 1760–1860," *Journal of the Early Republic* 36, no. 1 (2016): 37–72, and "Feeding Desire: Generative Environments, Meat Markets, and the Management of Sheep Intercourse in Great Britain, 1700–1750," *Osiris* 33, no. 1 (2018): 47–62; Gabriel N. Rosenberg, "No Scrubs: Livestock Breeding, Eugenics, and the State in the Early Twentieth-Century United States," *Journal of American History* 107, no. 2 (2020): 362–87; Harriet Ritvo, "Race, Breed, and Myths of Origin: Chillingham Cattle as Ancient Britons," *Representations* 39 (1992): 1–22, and *The Animal Estate: The English and Other Creatures in the Victorian Age* (Cambridge, MA: Harvard University Press, 1987).

10 Nigel Rothfels, "(Re)introducing the Przewalski's Horse," in *The Ark and Beyond: The Evolution of Zoo and Aquarium Conservation*, ed. Ben Minteer, Jane Maienschein, and James P. Collins (Chicago: University of Chicago Press: 2018), 79.

11 Lisa Onaga, "A Matter of Taste: Making Artificial Silkworm Food in Twentieth-Century Japan," in *Nature Remade: Engineering Life, Envisioning Worlds*, ed. by Luis A. Campos, Michael R. Dietrich, Tiago Saraiva, and Christian C. Young (Chicago: University of Chicago Press, 2021), 115–34.

12 Soraya de Chadarevian and Nick Hopwood, *Models: The Third Dimension of Science* (Stanford, CA: Stanford University Press, 2004); Manfred D. Laubichler and Gerd B. Müller. "Models in Theoretical Biology," *Modeling Biology: Structures, Behavior, Evolution*, eds. Manfred D. Laubichler and Gerd B. Müller (Cambridge, MA: MIT Press, 2007), 1–14.

13 There are several words that could be used in this chapter to describe this process. *Mariculture* is the process of rearing marine organisms specifically.

However, I've chosen to use the more blanket term of *aquaculture* for simplicity.

14 Jonathan Malindine, "Prehistoric Aquaculture: Origins, Implications, and an Argument for Inclusion," *Culture, Agriculture, Food and Environment* (2019): 66–70.

15 Tomoyoshi Komiyama, Hiroyuki Kobayashi, Yoshio Tateno, Hidetoshi Inoko, Takashi Gojobori, and Kazuho Ikeo, "An Evolutionary Origin and Selection Process of Goldfish," *Gene* 430, no. 1/2 (2009): 5–11; Lijing Jiang, "Retouching the Past with Living Things: Indigenous Species, Tradition, and Biological Research in Republican China, 1918–1937," *Historical Studies in the Natural Sciences* 46, no. 2 (2016): 154–206.

16 The Academy of Natural Sciences of Drexel University in Philadelphia has one of three copies of this manuscript in the United States. I was lucky enough to be able to view this work while at the Academy in 2005, and it is truly beautiful. See the online blog post "fantastic goldfish" from the Academy of Natural Sciences of Drexel University, https://ansp.org/exhibits/online-exhibits/stories/fantastic-goldfish/, and Edme Billardon de Sauvigny and François Nicolas Martinet, *Histoire naturelle des dorades de la Chine* (Paris: L'Imprimerie de Louis Jorry, 1780).

17 Lijing Jiang, "Crafting Socialist Embryology: Dialectics, Aquaculture and the Diverging Discipline in Maoist China, 1950–1965," *History and Philosophy of the Life Sciences* 40, no. 1 (2018): 3, "The Socialist Origins of Artificial Carp Reproduction in Maoist China," *Science, Technology and Society* 22, no. 1 (2017): 59–77, and "Retouching the Past."

18 Joseph E. Taylor III, *Making Salmon: An Environmental History of the Northwest Fisheries Crisis* (Seattle: University of Washington Press, 2009.

19 Livingstone Stone, "The Quinnat Salmon or California Salmon—*Oncorhynchus chouicha*," in *The Fisheries and Fishery Industries of the United States. Section I. Natural History of Useful Aquatic Animals*, prepared by George Brown Goode (Washington, DC: US Government Printing Office, 1884), 479–85, *Trout Culture* (Caledonia, NY: Press of Curtis, Morey, 1870), and "Trout Culture," *Transactions of the American Fisheries Society* 1, no. 1 (1872): 46–56; Sylvia R. Black, "Seth Green: Father of Fish Culture," *Rochester History* 6, no. 3 (1944): 9–10.

20 The commission on fish and fisheries was originally set up to be a temporary investigative group. After its establishment as a permanent organization, it was attached to the department of commerce. Eventually, it would split in two, with fresh water largely contained under the US Fish and Wildlife Service and marine stocks under NOAA.

21 Samantha K. Muka, "Working at Water's Edge: Life Sciences at American Marine Stations, 1880–1930" (PhD diss., University of Pennsylvania, 2014).

22 "Genetics Lab. Genetics and Correlated Studies of Fishes. N.D.T.S.," in Box 1 RG 7, Control Number 3009, Office of Director and Curators Breeder

and Coates 1939–1954, New York Zoological Society Archives, Wildlife Conservation Society, Bronx, NY; Xiphophorus Genetic Stock Center, Texas State University, https://www.xiphophorus.txstate.edu/about/introduction.html; Myron Gordon, "Genetics of Melanomas in Fishes V: The Reappearance of Ancestral Micromelanophores in Offspring of Parents Lacking These Cells," *Cancer Research* 1, no. 8 (1941): 656–59, and "The Genetics of Fish Diseases," *Transactions of the American Fisheries Society* 83, no. 1 (1954): 229–40.

23 Jason R. Meyers, "Zebrafish: Development of a Vertebrate Model Organism," *Current Protocols Essential Laboratory Techniques* 16, no. 1 (2018): e19; Hans W. Laale, "The Biology and Use of Zebrafish, *Brachydanio rerio* in Fisheries Research: A Literature Review," *Journal of Fish Biology* 10, no. 2 (1977): 121–73; Christian Lawrence, "The Husbandry of Zebrafish (*Danio rerio*): A Review," *Aquaculture* 269, no. 1–4 (2007): 1–20; Jiang, "Crafting Socialist Embryology," 3, and "Retouching the Past."

24 Crawford, D. R. Crawford, "Spawning Habits of the Spiny Lobster (*Panulirus argus*), with Notes on Artificial Hatching," *Transactions of the American Fisheries Society* 50, no. 1 (1921): 319.

25 Anna Marie Eleanor Roos, *Goldfish* (London: Reaktion Books, 2019).

26 William Thornton Innes, *Goldfish Varieties and Tropical Aquarium Fishes: A Complete Guide to Aquaria and Related Subjects* (Philadelphia: Innes and Sons, 1917).

27 F. H. Hoff, *Conditioning, Spawning and Rearing of Fish with Emphasis on Marine Clownfish*" Dade City, FL: Aquaculture Consultants, 1996), 1–6.

28 Martin A. Moe, Robert M. Ingle, and Raymond H. Lewis, *Pompano Mariculture: Preliminary Data and Basic Considerations* (St. Petersburg: Florida Board of Conservation Marine Laboratory, 1968), 1.

29 Ibid., 3.

30 E. M. Groover, J. van Senten, and M. Schwarz, "Species Profile: Clownfish," *USDA Southern Regional Aquaculture Center Publication*, no.7213 (2017), https://srac.tamu.edu/categories/view/33.

31 Martin Moe, "Breeding the Clownfish, *Amphiprion ocellaris*," *Salt Water Aquarium* 9, no. 2 (March/April, 1973); see also Martin Moe, "The Way We Were: 1973: Breeding the Clownfish, *Amphiprion ocellaris*," *Advanced Aquarist* 11 (February 8, 2012), https://www.advancedaquarist.com/2012/2/breeder.

32 Hoff, "Conditioning, Spawning and Rearing of Fish," 3.

33 Ibid., 1.

34 Ibid., 6.

35 Ibid., 68

36 Ibid., 108

37 Frank H. Hoff and Terry W. Snell, *Plankton Culture Manual* (Dade City, FL: Florida Aqua Farms, 1987).

38 Moe, *Breeding the Orchid Dottyback*, 64.

39 Ibid. 224

40 Ibid., 62.

41 Moe, "The Way We Were."

42 Hoff, "Conditioning, Spawning and Rearing of Fish," 57.

43 Moe, *Breeding the Orchid Dottyback*, 59.

44 Ibid. 86.

45 Keith M. Martin-Smith and Amanda C. J. Vincent, "Seahorse Declines in the Derwent Estuary, Tasmania in the Absence of Fishing Pressure," *Biological Conservation* 123, no. 4 (2005): 533–45; Elanor M. Bell, Jacqueline F. Lockyear, Jana M. McPherson, A. Dale Marsden, and Amanda C. J. Vincent. "First Field Studies of an Endangered South African Seahorse, *Hippocampus capensis*," *Environmental Biology of Fishes* 67, no. 1 (2003): 35–46.

46 Moreau, Marie-Annick, H. J. Hall, and A. C. J. Vincent. *Proceedings of the First International Workshop on the Management and Culture of Marine Species Used in Traditional Medicines, July 4–9, 1998, Cebu City, Philippines* (Montreal: Project Seahorse, McGill University, 2000), 5.

47 Nancy Kim Pham and Junda Lin, "The Effects of Different Feed Enrichments on Survivorship and Growth of Early Juvenile Longsnout Seahorse, *Hippocampus reidi*," *Journal of the World Aquaculture Society* 44, no. 3 (2013): 435–46.

48 Koldewey and Martin-Smith. "Global Review of Seahorse Aquaculture."

49 Ibid.

50 Pouil et al., "Aquaculture of Marine Ornamental Fish"; Jonathan A. Moorhead and Chaoshu Zeng, "Development of Captive Breeding Techniques for Marine Ornamental Fish: A Review," *Reviews in Fisheries Science* 18, no. 4 (2010): 315–43; Ike Olivotto, Miquel Planas, Nuno Simões, G. Joan Holt, Matteo Alessandro Avella, and Ricardo Calado, "Advances in Breeding and Rearing Marine Ornamentals," *Journal of the World Aquaculture Society* 42, no. 2 (2011): 135–66.

51 Thomas Ogawa and Christopher L. Brown, "Ornamental Reef Fish Aquaculture and Collection in Hawaii," *Aquarium Sciences and Conservation* 3, no. 1–3 (2001): 151–69;

52 Chatham K. Callan, Charles W. Laidley, Ian P. Forster, Kenneth M. Liu, Linda J. Kling, and Allen R. Place, "Examination of Broodstock Diet Effects on Egg Production and Egg Quality in Flame Angelfish (*Centropyge loriculus*)," *Aquaculture Research* 43, no. 5 (2012): 696–705; Charles W. Laidley, Chatham K. Callan, Andrew Burnell, K. M. Liu, Christina J. Bradley, M. Bou Mira, and Robin J. Shields, "Development of Aquaculture Technology for the Flame Angelfish (*Centropyge loriculus*)," *Regional Notes: Center for Tropical and Subtropical Aquaculture* 19, no. 2 (2008): 4–7.

53 Frank Baensch and Clyde S. Tamaru, "Spawning and Development of Larvae and Juveniles of the Rare Blue Mauritius Angelfish, *Centropyge debelius* (1988), in the Hatchery," *Journal of the World Aquaculture Society* 40, no. 4 (2009): 425–39; "Pygmy Angelfish Culture," Frank Baensch Fish Culture and Photography, https://www.frankbaensch.com/marine-aquarium-fish-culture/my-research/pygmy-angelfishes/culture/.

54 Chatham K. Callan, Aurora I. Burgess, Cara R. Rothe, and Renee Touse, "Development of Improved Feeding Methods in the Culture of Yellow Tang, *Zebrasoma flavescens*," *Journal of the World Aquaculture Society* 49, no. 3 (2018): 493–503; Aurora I. Burgess and Chatham K. Callan, "Effects of Supplemental Wild Zooplankton on Prey Preference, Mouth Gape, Osteological Development and Survival in First Feeding Cultured Larval Yellow Tang (*Zebrasoma flavescens*)," *Aquaculture* 495 (2018): 738–48. The Hawaii reefs have subsequently been closed to fishing.

55 Matthew A. DiMaggio, Eric J. Cassiano, Kevin P. Barden, Shane W. Ramee, Cortney L. Ohs, and Craig A. Watson, "First Record of Captive Larval Culture and Metamorphosis of the Pacific Blue Tang, *Paracanthurus hepatus*," *Journal of the World Aquaculture Society* 48, no. 3 (2017): 393–401.

56 Ibid. 399.

57 Ibid., 395.

58 Ibid., 399.

59 Ibid., 395.

60 Ibid. 400.

61 Eric J. Cassiano, Matthew L. Wittenrich, Thomas B. Waltzek, Natalie K. Steckler, Kevin P. Barden, and Craig A. Watson, "Utilizing Public Aquariums and Molecular Identification Techniques to Address the Larviculture Potential of Pacific Blue Tangs (*Paracanthurus hepatus*), Semicircle Angelfish (*Pomacanthus semicirculatus*), and Bannerfish (*Heniochus* sp.)," *Aquaculture International* 23, no. 1 (2015): 253–65.

62 Interview with the UF Tropical Aquarium Laboratory, December, 2017.

63 K. Madhu and Rema Madhu, "Captive Spawning and Embryonic Development of Marine Ornamental Purple Firefish *Nemateleotris decora* (Randall & Allen, 1973)," *Aquaculture* 424 (2014): 1–9; Kwee Siong Tew, Yun-Chen Chang, Pei-Jie Meng, Ming-Yih Leu, and David C. Glover, "Towards Sustainable Exhibits: Application of an Inorganic Fertilization Method in Coral Reef Fish Larviculture in an Aquarium," *Aquaculture Research* 47, no. 9 (2016): 2748–56; Ming-Yih Leu, Pei-Jie Meng, Chao-Sheng Huang, Kwee Siong Tew, Jimmy Kuo, and Chyng-Hwa Liou, "Spawning Behaviour, Early Development and First Feeding of the Bluestriped Angelfish [*Chaetodontoplus septentrionalis* (Temminck & Schlegel, 1844)] in Captivity," *Aquaculture Research* 41, no. 9 (2010): e39–e52; Ike Olivotto, Scott A. Holt, Oliana Carnevali, and G. Joan Holt, "Spawning, Early Development, and First

Feeding in the Lemonpeel Angelfish *Centropyge flavissimus*," *Aquaculture* 253, no. 1–4 (2006): 270–78.

Conclusion

1 The SeaWorld staff use the term *aquarist* for the work performed by guides in the cove. This includes leading swimmers, explaining what they are seeing, and also checking on the health and welfare of the animals in the enclosure.

2 Interview with Gary Violetta from Rising Tide and staff of Discovery Cove and SeaWorld, January 3, 2018.

3 Jean Baudrillard, *Simulacra and Simulation* (Ann Arbor: University of Michigan Press, 1994).

4 Rinkevich, Baruch. "Restoration strategies for coral reefs damaged by recreational activities: the use of sexual and asexual recruits." *Restoration Ecology* 3, no. 4 (1995): 241–251, and "Conservation of Coral Reefs through Active Restoration Measures: Recent Approaches and Last Decade Progress," *Environmental Science & Technology* 39, no. 12 (2005): 4333–42.

5 Madeleine J. H. van Oppen, James K. Oliver, Hollie M. Putnam, and Ruth D. Gates, "Building Coral Reef Resilience through Assisted Evolution," *Proceedings of the National Academy of Sciences* 112, no. 8 (2015): 2307–13.

6 I observed this scene on June 26, 2017, at the National Museum of Natural History while working as a postdoctoral fellow at the Smithsonian.

7 Interview with Valerie Paul, September 7, 2017

8 Aeby, Greta, Blake Ushijima, Justin E. Campbell, Scott Jones, Gareth Williams, Julie L. Meyer, Claudia Häse, and Valerie J. Paul, "Pathogenesis of a Tissue Loss Disease Affecting Multiple Species of Corals along the Florida Reef Tract," *Frontiers in Marine Science* 6 (2019): 678.

9 "World's First Coral Biobank in Port Douglas Key to Reef Survival," *Newport*, December. 17, 2019, https://www.newsport.com.au/2019/december/worlds-first-coral-biobank-in-port-douglas-key-to-reef-survival/

10 Nancy Knowlton, "Sea Urchin Recovery from Mass Mortality: New Hope for Caribbean Coral Reefs?," *Proceedings of the National Academy of Sciences* 98, no. 9 (2001): 4822–24; Dannise V. Ruiz-Ramos, Edwin A. Hernández-Delgado, and Nikolaos V. Schizas, "Population Status of the Long-Spined Urchin *Diadema antillarum* in Puerto Rico 20 Years after a Mass Mortality Event," *Bulletin of Marine Science* 87, no. 1 (2011): 113–27.

11 Aaron R. Pilnick, Keri L. O'Neil, Martin Moe, and Joshua T. Patterson, "A Novel System for Intensive *Diadema antillarum* Propagation as a Step towards Population Enhancement," *Scientific Reports* 11, no. 1 (2021): 1–13.

12 Juan M. Ezcurra, Christopher G. Lowe, Henry F. Mollet, Lara A. Ferry, and John B. O'Sullivan, "Captive Feeding and Growth of Young-of-the-Year White

Sharks, *Carcharodon carcharias*, at the Monterey Bay Aquarium," *Global Perspectives on the Biology and Life History of the White Shark*, ed. Michael L. Domeier (Boca Raton, FL: Taylor and Francis, 2012).

13 Bruce H. Robison, Kim R. Reisenbichler, James C. Hunt, and Steven H. D. Haddock, "Light Production by the Arm Tips of the Deep-Sea Cephalopod *Vampyroteuthis infernalis*," *Biological Bulletin* 205, no. 2 (2003): 102–9.

Bibliography

Adams, Jake. "Historic *Stuber acropora* Photograph Documents 25 Years of Stony Coral Reefing." Reef Builders, News, January 5, 2011. https://reefbuilders.com/2011/01/05/historic-stuber-acropora-photograph-documents-25-years-stony-coral-reefing/.

Adey, Walter H. "The Microcosm: A New Tool for Reef Research." *Coral Reefs* 1, no. 3 (1983): 193–201.

Adey, Walter H., and Karen Loveland. *Dynamic Aquaria: Building Living Ecosystems*. San Diego, CA: Academic Press, 1998.

Aeby, Greta S., Blake Ushijima, Justin E. Campbell, Scott Jones, Gareth J. Williams, Julie L. Meyer, Claudia Häse, and Valerie J. Paul. "Pathogenesis of a Tissue Loss Disease Affecting Multiple Species of Corals along the Florida Reef Tract." *Frontiers in Marine Science* 6 (2019): 678.

Allen, David. "23 Tastes and Crazes." In *Cultures of Natural History*, edited by N. Jardine, J. A. Secord, and E. C. Spary, 394–407. Cambridge: Cambridge University Press, 1996.

Allen, John, and Mark Nelson. "Overview and Design Biospherics and Biosphere 2, Mission One (1991–1993)." *Ecological Engineering* 13, nos. 1–4 (1999): 15–29.

Allmon, Warren D. "The Evolution of Accuracy in Natural History Illustration: Reversal of Printed Illustrations of Snails and Crabs in Pre-Linnaean Works Suggests Indifference to Morphological Detail." *Archives of Natural History* 34, no. 1 (2007): 174–91.

Anderson, Katharine. "Coral Jewelry." *Victorian Review* 34, no. 1 (2008): 47–52.

Andon, Alex. "Jellyfish Tank by Jellyfish Art." Kickstarter.com. https://www.kickstarter.com/projects/jellyfishart/jellyfish%ADaquarium/description (accessed May 17, 2017; no longer posted).

Armstrong, Isobel. *Victorian Glassworlds: Glass Culture and the Imagination 1830–1880*. Oxford: Oxford University Press, 2008.

Baensch, Frank, and Clyde S. Tamaru. "Spawning and Development of Larvae and Juveniles of the Rare Blue Mauritius Angelfish, *Centropyge debelius* (1988), in the Hatchery." *Journal of the World Aquaculture Society* 40, no. 4 (2009): 425–39.

Baker, L. D., and M. R. Reeve. "Laboratory Culture of the Lobate Ctenophore *Mnemiopsis mccradyi* with Notes on Feeding and Fecundity." *Marine Biology* 26, no. 1 (1974): 57–62.

Barnett, Tim P., David W. Pierce, and Reiner Schnur. "Detection of Anthropogenic Climate Change in the World's Oceans." *Science* 292, no. 5515 (2001): 270–74.

Bartlett, Thomas C. "Small Scale Experimental Systems for Coral Research: Considerations, Planning, and Recommendations." *NOAA Technical Memorandum NOS NCCOS* 165, 2013.

Baudrillard, Jean. *Simulacra and Simulation*. Ann Arbor: University of Michigan Press, 1994.

Bauer, Susanne, Martina Schlünder, and Maria Rentetzi, eds. *Boxes: A Field Guide*. Manchester: Mattering Press, 2020. https://www.matteringpress.org/books/boxes.

Beebe, William. *Half Mile Down*. New York: Duell, Sloan and Pearce, 1951.

Benson, Etienne S. *Surroundings: A History of Environments and Environmentalisms*. Chicago: University of Chicago Press, 2020.

Bell, Elanor M., Jacqueline F. Lockyear, Jana M. McPherson, A. Dale Marsden, and Amanda C. J. Vincent. "First Field Studies of an Endangered South African Seahorse, *Hippocampus capensis*." *Environmental Biology of Fishes* 67, no. 1 (2003): 35–46.

Bijma, Jelle, Hans-O. Pörtner, Chris Yesson, and Alex D. Rogers. "Climate Change and the Oceans–What Does the Future Hold?" *Marine Pollution Bulletin* 74, no. 2 (2013): 495–505.

Billardon de Sauvigny, Edme, and François Nicolas Martinet. *Histoire naturelle des dorades de la Chine*. Paris: L'Imprimerie de Louis Jorry, 1780.

Black, Sylvia R. "Seth Green: Father of Fish Culture." *Rochester History* 6, no. 3 (1944): 9–10.

Blum, Ann Shelby. *Picturing Nature: American Nineteenth-Century Zoological Illustration* Princeton, NJ: Princeton University Press, 1993.

Bousé, Derek. *Wildlife Films*. Philadelphia: University of Pennsylvania Press, 2011. Paraphrased in Eleanor Louson, "Taking Spectacle Seriously: Wildlife Film and the Legacy of Natural History Display," *Science in Context* 31, no. 1 (2018): 16.

Bowen, James. *The Coral Reef Era: From Discovery to Decline; A History of Scientific Investigation from 1600 to the Anthropocene Epoch*. New York: Springer, 2015.

Boyd, William. "Making Meat: Science, Technology, and American Poultry Production." *Technology and Culture* 42, no. 4 (2001): 631–64.

Boyle, Robert H. "The Strange Fish and Stranger Times of Dr. Herbert R. Axelrod." *Sports Illustrated*. May 3, 1965.

Browne, Edward T. "On Keeping Medusae Alive in an Aquarium." *Journal of the Marine Biological Association of the United Kingdom* 5, no. 2 (1898): 176–80.

Brunner, Bernd. *The Ocean at Home: An Illustrated History of the Aquarium*. London: Reaktion Books, 2012.

Bruno, John F., Elizabeth R. Selig, Kenneth S. Casey, Cathie A. Page, Bette L.

Willis, C. Drew Harvell, Hugh Sweatman, and Amy M. Melendy. "Thermal Stress and Coral Cover as Drivers of Coral Disease Outbreaks." *PLoS Biology* 5, no. 6 (2007): e124. https://dx.doi.org/10.1371%2Fjournal.pbio.0050124.

Burgess, Aurora I., and Chatham K. Callan. "Effects of Supplemental Wild Zooplankton on Prey Preference, Mouth Gape, Osteological Development and Survival in First Feeding Cultured Larval Yellow Tang (*Zebrasoma flavescens*)." *Aquaculture* 495 (2018): 738–48.

Caballes, Ciemon Frank, and Morgan S. Pratchett. "Environmental and Biological Cues for Spawning in the Crown-of-Thorns Starfish." *PLoS* 12, no. 3, (2017): e0173964. https://www.ncbi.nlm.nih.gov/pmc/articles/PMC5371309/.

Callan, Chatham K., Aurora I. Burgess, Cara R. Rothe, and Renee Touse. "Development of Improved Feeding Methods in the Culture of Yellow Tang, *Zebrasoma flavescens*." *Journal of the World Aquaculture Society* 49, no. 3 (2018): 493–503.

Callan, Chatham K., Charles W. Laidley, Ian P. Forster, Kenneth M. Liu, Linda J. Kling, and Allen R. Place. "Examination of Broodstock Diet Effects on Egg Production and Egg Quality in Flame Angelfish (*Centropyge loriculus*)." *Aquaculture Research* 43, no. 5 (2012): 696–705.

Campos, Luis A., Michael R. Dietrich, Tiago Saraiva, and Christian C. Young, eds. *Nature Remade: Engineering Life, Envisioning Worlds*. Chicago: University of Chicago Press, 2021.

Canino, Michael F., Ingrid B. Spies, Kathryn M. Cunningham, Lorenz Hauser, and W. Stewart Grant. "Multiple Ice-Age Refugia in Pacific Cod, *Gadus macrocephalus*." *Molecular Ecology* 19, no. 19 (2010): 4339–51.

Carlson, Bruce A. "Aquarium Systems for Living Corals." *International Zoo Yearbook* 26, no.1 (1987): 1–9.

———. "General Introduction: Advances in Coral Husbandry in Public Aquaria." In *Advances in Coral Husbandry in Public Aquariums*, Public Aquarium Husbandry Series, vol. 2, edited by R. J. Leewis and M. Janse, ix–xv. Arnhem, The Netherlands: Burgers' Zoo, 2008.

———. "Organism Responses to Rapid Change: What Aquaria Tell Us About Nature." *American Zoologist* 39 no. 1 (1999): 45–55.

Cassiano, Eric J., Matthew L. Wittenrich, Thomas B. Waltzek, Natalie K. Steckler, Kevin P. Barden, and Craig A. Watson. "Utilizing Public Aquariums and Molecular Identification Techniques to Address the Larviculture Potential of Pacific Blue Tangs (*Paracanthurus hepatus*), Semicircle Angelfish (*Pomacanthus semicirculatus*), and Bannerfish (*Heniochus* sp.)." *Aquaculture International* 23, no. 1 (2015): 253–65.

Catala, René. *Carnival under the Sea*. Paris: R. Sicard, 1964.

Clarke, Adele E., and Joan H. Fujimura, eds. *The Right Tools for the Job: At Work in Twentieth-Century Life Sciences*. Princeton, NJ: Princeton University Press, 2014.

Cohen, Margaret. *The Novel and the Sea*. Princeton, NJ: Princeton University Press, 2021.

———. "The Underwater Imagination: From Environment to Film Set, 1954–1956." *English Language Notes* 57, no. 1 (2019): 51–71.

Coles, S. L., and Paul L. Jokiel. "Synergistic Effects of Temperature, Salinity and Light on the Hermatypic Coral *Montipora verrucosa*." *Marine Biology* 49, no. 3 (1978): 187–95.

Coles, Stephen L., Paul L. Jokiel, and Clark R. Lewis. "Thermal Tolerance in Tropical versus Subtropical Pacific Reef Corals." *Pacific Science* 30, no. 2 (1976): 159–66.

Colin, Patrick L. "A Brief History of the Tortugas Marine Laboratory and the Department of Marine Biology, Carnegie Institution of Washington." In *Oceanography: The Past*, edited by M. Sears and D. Merriman, 138–47. New York: Springer, 1980.

Collins, Harry M. "Tacit Knowledge, Trust and the Q of Sapphire." *Social Studies of Science* 31, no. 1 (2001): 71–85.

———. "What Is Tacit Knowledge?" In *The Practice Turn in Contemporary Theory*, edited by Theodore R. Schatzki, 115–28. London: Routledge, 2005.

Collins, Harry M., and Robert Evans. *Rethinking Expertise*. Chicago: University of Chicago Press, 2008.

Cooke, Kathy J. "From Science to Practice, or Practice to Science? Chickens and Eggs in Raymond Pearl's Agricultural Breeding Research, 1907–1916." *Isis* 88, no. 1 (1997): 62–86.

Cooper, Isabel. "Artist at Large." *Atlantic Monthly*, July 1926, 85–93.

Corbin, Alain. *The Lure of the Sea: The Discovery of the Seaside in the Western World, 1750–1840*. Berkeley: University of California Press, 1994.

Crawford, D. R. "Spawning Habits of the Spiny Lobster (*Panulirus argus*), with Notes on Artificial Hatching." *Transactions of the American Fisheries Society* 50, no. 1 (1921): 312–19.

Crylen, Jonathan Christopher. "The Cinematic Aquarium: A History of Undersea Film." PhD diss., University of Iowa, 2015.

Daston, Lorraine, and Peter Galison. *Objectivity*. Princeton, NJ: Princeton University Press, 2021.

David, P. M. "The Photography of Live Oceanic Plankton Animals." *International Photo Tecknik* (1963): 40–43.

"A Decade of Discovery." Census of Marine Life, 2010. http://www.coml.org/.

de Chadarevian, Soraya, and Nick Hopwood. *Models: The Third Dimension of Science*. Stanford, CA: Stanford University Press, 2004.

Delap, Maude J. "Notes on the Rearing, in an Aquarium, of *Aurelia aurita*, L. and *Pelagia perla* (Slabber)." *Report on the Sea and Inland Fisheries of Ireland for 1905*, 160–64 and 2 plates. Dublin: n.p., 1907.

———. "Notes on the Rearing, in an Aquarium, of *Cyanea lamarcki*, Péron et Lesueur." In *Report on the Sea and Island Fisheries of Ireland for 1902–03*, 20–22. Dublin: n.p., 1906.

———. "Notes on the Rearing of *Chrysaora isosceles* in an Aquarium." *Irish Naturalist* 10, no. 2 (February 1901): 25–28.

Delrieu-Trottin, Erwan, Jeffrey T. Williams, and Serge Planes. "*Macropharyngodon pakoko*, a New Species of Wrasse (Teleostei: Labridae) Endemic to the Marquesas Islands, French Polynesia." *Zootaxa* 3857, no. 3 (2014): 433–43.

DiMaggio, Matthew A., Eric J. Cassiano, Kevin P. Barden, Shane W. Ramee, Cortney L. Ohs, and Craig A. Watson. "First Record of Captive Larval Culture and Metamorphosis of the Pacific Blue Tang, *Paracanthurus hepatus*." *Journal of the World Aquaculture Society* 48, no. 3 (2017): 393–401.

DiMaggio, Matthew A., E. M. Groover, J. van Senten, and M. Schwarz. "Species Profile: Clownfish." *USDA Southern Regional Aquaculture Center Publication*, no.7213 (2017). https://srac.tamu.edu/categories/view/33.

Drum and Croaker: A Highly Irregular Journal for the Public Aquarist. http://drumandcroaker.org/.

Duffy, Rebecca. "The Age of Aquaria: The Aquarium Pursuit and Personal Fish-Keeping, 1850–1920." PhD diss., University of Delaware, 2018.

Dunton, Sam C. *Guide to Photographing Animals*. New York: Greenberg, 1956.

Edmonds, Samantha. "The Miraculous Journey of a Captive-Bred Hermit Crab." *Outline*, October 18, 2019. https://theoutline.com/post/8116/hermit-crabs-breeding-captivity?zd=1&zi=6qeiqbye.

Eigen, Edward. "Dark Space and the Early Days of Photography as a Medium." *Grey Room* 3 (2001): 90–111.

———. "On the Screen and in the Water: On Photographically Envisioning the Sea." In *L'architecture, les sciences et la culture de l'histoire au XIXe siècle*, 229–48. Saint-Etienne: Publications de l'Université de Saint-Etienne, 2001.

Elias, Ann. *Coral Empire: Underwater Oceans, Colonial Tropics, Visual Modernity*. Durham, NC: Duke University Press, 2019.

Emery, Alan R., and Richard Winterbottom. "A Technique for Fish Specimen Photography in the Field." *Canadian Journal of Zoology* 58, no. 11 (1980): 2158–62.

Endt-Jones, Marion. "'Something Rich and Strange': Coral in Contemporary Art." *Framing the Ocean, 1700 to the Present: Envisaging the Sea as Social Space*, edited by Tricia Cusack, 223–38. London: Routledge, 2017.

Eng, Lee Chin. "Nature's System of Keeping Marine Fishes." *Tropical Fish Hobbyist* 9, no. 6 (1961): 23–30.

Evans, Georgina. "Framing Aquatic Life." *Screen* 61, no. 2 (2020): 170.

Ezcurra, Juan M., Christopher G. Lowe, Henry F. Mollet, Lara A. Ferry, and John B. O'Sullivan. "Captive Feeding and Growth of Young-of-the-Year White Sharks, *Carcharodon carcharias*, at the Monterey Bay Aquarium." In *Global Perspectives on the Biology and Life History of the White Shark*, edited by Michael L. Domeier, 3–15. Boca Raton, FL: Taylor and Francis, 2012

Fabian, Ann. *The Skull Collectors: Race, Science, and America's Unburied Dead*. Chicago: University of Chicago Press, 2010.

Fabre-Domergue, P. *La photographie des animaux aquatiques*. Paris: George Carré et C. Naud, 1899.

Fabre-Domergue, Paul Louis Marie, and Eugène Biétrix. *Développement de la sole*

(Solea vulgaris*): Introduction à l'étude de la pisciculture marine*. Paris: Vuibert and Nony, 1905.

Farber, Paul Lawrence. *Finding Order in Nature: The Naturalist Tradition from Linnaeus to E. O. Wilson*. Baltimore: Johns Hopkins University Press, 2000.

"The Father of Modern Reef Keeping: Lee Chin Eng." *Reef Aquarium Farming News*, no. 6 (June 1997): 3. http://www.garf.org/news6p3.html.

"Filming the Impossible Sets: Filming Burrows and Tanks." BBC News, April 29, 2016. http://www.bbc.com/earth/story/20160310-filming-the-impossible-sets-filming-burrows-and-tanks (accessed December 12, 2019; no longer posted).

Finley, Carmel. *All the Fish in the Sea: Maximum Sustainable Yield and the Failure of Fisheries Management*. Chicago: University of Chicago Press, 2011.

Florida Aquarium. "The Florida Aquarium Becomes the First Organization in History to Induce Spawning of Atlantic Coral." August 21, 2019. https://floridascoralreef.org/2019/08/21/the-florida-aquarium-becomes-first-organization-in-history-to-induce-spawning-of-atlantic-coral-a-new-hope-to-save-floridas-reefs.

Franz, Kathleen. *Tinkering: Consumers Reinvent the Early Automobile*. Philadelphia: University of Pennsylvania Press, 2011.

Frehill, Lisa M. "The Gendered Construction of the Engineering Profession in the United States, 1893–1920." *Men and Masculinities* 6, no. 4 (2004): 383–403.

Fujimura, Joan H. "Standardizing Practices: A Socio-History of Experimental Systems in Classical Genetic and Virological Cancer Research, ca. 1920–1978." *History and Philosophy of the Life Sciences* 18, no. 1 (1996): 3–54.

Fyfe, Aileen. *Science and Salvation: Evangelical Popular Science Publishing in Victorian Britain*. Chicago: University of Chicago Press, 2004.

Gershwin, Lisa-Ann. *Stung! On Jellyfish Blooms and the Future of the Ocean*. Chicago: University of Chicago Press, 2013.

Gershwin, Lisa-Ann, and Allen G. Collins. "A Preliminary Phylogeny of Pelagiidae (Cnidaria, Scyphozoa), with new observations of *Chrysaora colorata* comb. nov." *Journal of Natural History* 36, no. 2 (2002): 127–48.

Gignoux-Wolfsohn, S. A., Christopher J. Marks, and Steven V. Vollmer. "White Band Disease Transmission in the Threatened Coral, *Acropora cervicornis*." *Scientific Reports* 2 (2012): 1–3.

Gilchrist, F. G. "Rearing the Scyphistoma of *Aurelia* in the Laboratory." In *Culture Methods for Invertebrate Animals: A Compendium Prepared Cooperatively by American Zoologists under the Direction of a Committee from Section F of the American Association for the Advancement of Science*, edited by Frank E. Lutz, Paul L. Welch, and Paul S. Galtsoff, 143–44. New York: Dover, 1937.

Goldstein, Jason S., and Brian Nelson. "Application of a Gelatinous Zooplankton Tank for the Mass Production of Larval Caribbean Spiny Lobster, *Panulirus argus*." *Aquatic Living Resources* 24, no. 1 (2011): 45–51.

Goodbred, Steven, and Thomas Occhiogrosso. "Method for Photographing Small Fish." *Progressive Fish-Culturist* 41, no. 2 (1979): 76–77.

Gordon, Myron. "The Genetics of Fish Diseases." *Transactions of the American Fisheries Society* 83, no. 1 (1954): 229–40.

Gordon, Myron. "Genetics of Melanomas in Fishes V: The Reappearance of Ancestral Micromelanophores in Offspring of Parents Lacking These Cells." *Cancer Research* 1, no. 8 (1941): 656–59.

Granata, Silvia. "'At Once Pet, Ornament, and "Subject for Dissection"': The Unstable Status of Marine Animals in Victorian Aquaria." *Cahiers victoriens et édouardiens* 88 (Autumn 2018).

———. "The Dark Side of the Tank: The Marine Aquarium in the Victorian Home." In *Paraphernalia! Victorian Objects*, edited by Helen Kingstone and Kate Lister, 81–98. London: Routledge, 2018.

———. "'Let Us Hasten to the Beach': Victorian Tourism and Seaside Collecting." *Lit: Literature Interpretation Theory* 27, no. 2 (2016): 91–110.

———. "The Victorian Aquarium as a Miniature Sea." *Underwater Worlds: Submerged Visions in Science and Culture* 19, no. 1 (2019): 108–28.

Greene, Mott. "Arctic Sea Ice, Oceanography, and Climate Models." In *Extremes: Oceanography's Adventures at the Poles*, edited by Keith R. Benson and Helen M. Rozwadowski, 305–12. Sagamore Beach, MA: Science History, 2007.

Green, Seth. *Trout Culture*. Rochester, NY: Press of Curtis, Morey, 1870.

Greene, Jeremy A. "Do-It-Yourself Medical Devices: Technology and Empowerment in American Health Care." *New England Journal of Medicine* 374, no. 4 (2016): 305–8.

Greenfield, D. W., and J. E. Randall. "*Myersina balteata*, a New Shrimp-Associated Goby (Teleostei: Gobiidae) from Guadalcanal, Solomon Islands." *Journal of the Ocean Science Foundation* 30 (2018): 90–99.

Greve, Wulf. "Cultivation Experiments on North Sea Ctenophores." *Helgoländer wissenschaftliche Meeresuntersuchungen* 20, no. 1 (1970): 304–10.

———. "The 'Meteor Planktonküvette': A Device for the Maintenance of Macrozooplankton Aboard Ships." *Aquaculture* 6, no. 1 (1975): 77–82.

———. "The 'Planktonkreisel': A New Device for Culturing Zooplankton." *Marine Biology* 1, no. 3 (1968): 201–3.

Hamera, Judith. *Parlor Ponds: The Cultural Work of the American Home Aquarium, 1850–1970*. Ann Arbor: University of Michigan Press, 2012.

Hamlin, Christopher. "Robert Warington and the Moral Economy of the Aquarium." *Journal of the History of Biology* (1986): 131–53.

Hamner, William M. "Design Developments in the Planktonkreisel, a Plankton Aquarium for Ships at Sea." *Journal of Plankton Research* 12, no. 2 (1990): 397–402.

Hanson, Elizabeth. *Animal Attractions: Nature on Display in American Zoos*. Princeton, NJ: Princeton University Press, 2004.

Haraway, Donna J. *Modest_Witness@Second_Millennium: FemaleMan_Meets_OncoMouse: Feminism and Technoscience*. With a preface and study guide by Thyrza Nichols Goodeve. New York: Routledge, 2018.

Harrould-Kolieb, Ellycia R., and Dorothée Herr. "Ocean Acidification and

Climate Change: Synergies and Challenges of Addressing Both under the UNFCCC." *Climate Policy* 12, no. 3 (2012): 378–89.

Hayman, Emma, and Peter Hannam. "'Coral Ark' Bids to Save Biodiversity as Global Threats to Reefs Mount." *Sydney Morning Herald*, December 15, 2019. https://www.smh.com.au/environment/sustainability/coral-ark-bids-to-save-biodiversity-as-global-threats-to-reefs-mount-20191213-p53jw8.html?fbclid=IwAR1xavP8O4jOw3dm22lWu3g9JKow-4gOH8nXcrNfN_Ad0evuWF4ai8_Ox0k.

Hayward, Eva. "Sensational Jellyfish: Aquarium Affects and the Matter of Immersion." *differences* 23, no. 3 (2012): 161–96.

Helm, Rebecca R. "New Research Reveals How to Easily Grow Jellyfish in Captivity." *Deep Sea News*, December 28, 2017. https://www.deepseanews.com/2017/12/new-research-reveals-how-to-easily-grow-jellyfish-in-captivity/.

Helm, Rebecca R., and Casey W. Dunn. "Indoles Induce Metamorphosis in a Broad Diversity of Jellyfish, but Not in a Crown Jelly (Coronatae)." *PLoS One* 12, no. 12 (2017): https://doi.org/10.1371/journal.pone.0188601.

Herler, Juergen, Lovrenc Lipej, and Tihomir Makovec. "A Simple Technique for Digital Imaging of Live and Preserved Small Fish Specimens." *Cybium* 31, no. 1 (2007): 39–44.

Hershey, David R. "Doctor Ward's Accidental Terrarium." *American Biology Teacher* 58, no. 5 (1996): 276–81.

Higginbotham, James Arnold. *Piscinae: Artificial Fishponds in Roman Italy.* Chapel Hill: University of North Carolina Press, 1997.

Hirai, Etur. "On the Developmental Cycles of *Aurelia aurita* and *Doctylometra pacifica*." *Bulletin of the Marine Biological Station of Asamushi, Tohoku University* 9 (1958): 81.

———. "On the Species of *Cladonema radiatum* var. *mayeri* Perkins." *Bulletin of the Marine Biological Station of Asamushi, Tohoku University* 9 (1958): 23–25.

Hirota, Jed. "Laboratory Culture and Metabolism of the Planktonic Ctenophore, *Pleurobrachia bachei* A. Agassiz." In *Biological Oceanography of the Northern North Pacific Ocean*, edited by A. Y. Takenouti, 465. Tokyo: Idemitsu Shoten, 1972.

Hoegh-Guldberg, Ove, Elvira S. Poloczanska, William Skirving, and Sophie Dove. "Coral Reef Ecosystems under Climate Change and Ocean Acidification." *Frontiers in Marine Science* 4 (2017). https://doi.org/10.3389/fmars.2017.00158.

Hoff, Frank H. *Conditioning, Spawning and Rearing of Fish with Emphasis on Marine Clownfish.* Dade City, FL: Aquaculture Consultants, 1996.

Hoff, Frank H., and Terry W. Snell. *Plankton Culture Manual.* Dade City, FL: Florida Aqua Farms, 1987.

Holm, Erling. "Improved Technique for Fish Specimen Photography in the Field." *Canadian Journal of Zoology* 67, no. 9 (1989): 2329–32.

Howe, Jeffrey C. "A Technique for Immobilizing and Photographing Small, Live Fishes." *Fisheries Research* 27, no. 4 (1996): 261–64.

Hubbard, J. M. *A Science on the Scales: The Rise of Canadian Atlantic Fisheries Biology, 1898–1939*. Toronto: University of Toronto Press, 2006.

Hugo, Victor. *Toilers of the Sea*. Translated by W. Moy Thomas and illustrated by Gustave Doré. London: Sampson Low, Son, and Marston, 1867.

Humblefish. "A Tribute to Lee Chin Eng!" Nano-Reef.com, April 21, 2020. https://www.nano-reef.com/forums/topic/412435-a-tribute-to-lee-chin-eng/.

Innes, William Thornton. *Goldfish Varieties and Tropical Aquarium Fishes: A Complete Guide to Aquaria and Related Subjects*. Philadelphia: Innes and Sons, 1917.

———. "Aquarium Fish Photography." *Complete Photographer* 1, no. 4 (1941): 238–48.

Jackson, Christine E. "The Materials and Methods of Hand-Colouring Zoological Illustrations." *Archives of Natural History* 38, no.1 (2011): 53–64.

———. "The Painting of Hand-Coloured Zoological Illustrations." *Archives of Natural History* 38, no. 1 (2011): 36–52.

Jacob, François. "Evolution and Tinkering." *Science* 196, no. 4295 (1977): 1161–66.

Jaubert, J. "An Integrated Nitrifying-Denitrifying Biological System Capable of Purifying Sea Water in a Closed Circuit Aquarium." *Bulletin de l'Institut. Océanographique, Monaco* 5 (1989): 101–6.

———. "Scientific Considerations on a Technique of Ecological Purification That Made Possible the Cultivation of Reef-Building Corals in Monaco." In *Advances in Coral Husbandry in Public Aquariums*, edited by Rob J. Leewis and Max Janse, 115–26. Arnhem, The Netherlands: Burgers' Zoo, 2008.

"Jellyfishart/EON/Cubic Tank Discussion and Fan Page." https://www.facebook.com/groups/397225970290084/search/?query=problem.

JellyfishArtDotCom. "Jellyfish Aquarium Kickstarter w Vanilla Ice." 2017. https://www.youtube.com/watch?v=7LbC7_ZhrJQ.

Jiang, Lijing. "Crafting Socialist Embryology: Dialectics, Aquaculture and the Diverging Discipline in Maoist China, 1950–1965." *History and Philosophy of the Life Sciences* 40, no. 1 (2018): 3.

———. "Retouching the Past with Living Things: Indigenous Species, Tradition, and Biological Research in Republican China, 1918–1937." *Historical Studies in the Natural Sciences* 46, no. 2 (2016): 154–206.

———. "The Socialist Origins of Artificial Carp Reproduction in Maoist China." *Science, Technology and Society* 22, no. 1 (2017): 59–77.

Jokiel, Paul L. "Solar Ultraviolet Radiation and Coral Reef Epifauna." *Science* 207, no. 4435 (1980): 1069–71.

Jørgensen, Dolly. "Mixing Oil and Water: Naturalizing Offshore Oil Platforms in Gulf Coast Aquariums." *Journal of American Studies* 46, no. 2 (2012): 461–80.

Jørgensen, Dolly, Finn Arne Jørgensen, and Sara B. Pritchard, eds. *New Natures: Joining Environmental History with Science and Technology Studies*. Pittsburgh, PA: University of Pittsburgh Press, 2013.

Kakinuma, Y. "An Experimental Study of the Life Cycle and Organ Differentiation of *Aurelia aurita* Lamarck." *Bulletin of the Marine Biological Station of Asamushi, Tohoku University* 15 (1975): 101–12.

Kamshilov, M. M. "The Dependence of Ctenophore *Beroe cucumis* Fab Sizes from Feeding" [in Russian]. *Doklady of the Academy of Sciences of the USSR* 131 (1960): 957–60.

———. "Pitanie Grebnevika Beroe-Cucumis Fab." *Doklady of the Academy of Sciences of the USSR* 102, no. 2 (1955): 399–402.

Keeney, Elizabeth. *The Botanizers: Amateur Scientists in Nineteenth-Century America*. Chapel Hill: University of North Carolina Press, 1992.

Keiner, Christine. "Modeling Neptune's Garden: The Chesapeake Bay Hydraulic Model, 1965–1984." In *The Machine in Neptune's Garden: Historical Perspectives on Technology and the Marine Environment*, edited by Helen M. Rozwadowski and David K. Van Keuren, 273–314. Sagamore Beach, MA: Science History, 2004.

Kemp, S., and A. V. Hill. "Edgar Johnson Allen. 1866–1942." *Obituary Notices of Fellows of the Royal Society* 4, no. 12 (1943): 361.

Kimura, Aya H., and Abby Kinchy. "Citizen Science: Probing the Virtues and Contexts of Participatory Research." *Engaging Science, Technology, and Society* 2 (2016): 331–61.

Kisling, Vernon N. "Historic and Cultural Foundations of Zoo Conservation: A Narrative Timeline." In *The Ark and Beyond: The Evolution of Aquarium and Zoo Conservation*, edited by Ben A. Minteer, Jane Maienschein, and James P. Collins, 41–50. Chicago: University of Chicago Press, 2018.

———. *Zoo and Aquarium History: Ancient Animal Collections to Zoological Gardens*. Boca Raton, FL: CRC, 2000.

Knight, David. *Zoological Illustration: An Essay towards a History of Printed Zoological Pictures*. Folkstone, Kent: Wm Dawson and Son, 1977.

Knorr-Cetina, Karen. *Epistemic Cultures: How Science Makes Sense*. N.p.: n.p., 1995.

Knowlton, Nancy. "Sea Urchin Recovery from Mass Mortality: New Hope for Caribbean Coral Reefs?" *Proceedings of the National Academy of Sciences* 98, no. 9 (2001): 4822–24.

Kohler, Robert E. *Lords of the Fly: Drosophila Genetics and the Experimental Life*. Chicago: University of Chicago Press, 1994.

Koldewey, Heather J., and Keith M. Martin-Smith. "A Global Review of Seahorse Aquaculture." *Aquaculture* 302, no. 3/4 (2010): 131–52.

Komiyama, Tomoyoshi, Hiroyuki Kobayashi, Yoshio Tateno, Hidetoshi Inoko, Takashi Gojobori, and Kazuho Ikeo. "An Evolutionary Origin and Selection Process of Goldfish." *Gene* 430, no. 1/2 (2009): 5–11.

Kooperman, Netta, Eitan Ben-Dov, Esti Kramarsky-Winter, Zeev Barak, and Ariel Kushmaro. "Coral Mucus-Associated Bacterial Communities from Natural and Aquarium Environments." *FEMS Microbiology Letters* 276, no. 1 (2007): 106–13.

Kuffner, Ilsa B., Linda J. Walters, Mikel A. Becerro, Valerie J. Paul, Raphael

Ritson-Williams, and Kevin S. Beach. "Inhibition of Coral Recruitment by Macroalgae and Cyanobacteria." *Marine Ecology Progress Series* 323 (2006): 107–17.

Laale, Hans W. "The Biology and Use of Zebrafish, *Brachydanio rerio* in Fisheries Research: A Literature Review." *Journal of Fish Biology* 10, no. 2 (1977): 121–73.

Laidley, Charles W., Chatham K. Callan, Andrew Burnell, K. M. Liu, Christina J. Bradley, M. Bou Mira, and Robin J. Shields. "Development of Aquaculture Technology for the Flame Angelfish (*Centropyge loriculus*)." *Regional Notes: Center for Tropical and Subtropical Aquaculture* 19, no. 2 (2008): 4–7.

Lange, Jürgen, and Rainer Kaiser. "The Maintenance of Pelagic Jellyfish in the Zoo-Aquarium Berlin." *International Zoo Yearbook* 34, no. 1 (1995): 59–64.

Larson, Roy "Obituary: Charles E. Cuttress, 1921–1992." *Bulletin of Marine Sciences* 51, no. 3 (1992): 480–81.

Latour, Bruno. *Science in Action: How to Follow Scientists and Engineers through Society*. Cambridge, MA: Harvard University Press, 1987.

Laubichler, Manfred D., and Gerd B. Müller. "Models in Theoretical Biology." *Modeling Biology: Structures, Behavior, Evolution*, edited by Manfred D. Laubichler and Gerd B. Müller, 1–14. Cambridge, MA: MIT Press, 2007.

Lawrence, Christian. "The Husbandry of Zebrafish (*Danio rerio*): A Review." *Aquaculture* 269, no. 1–4 (2007): 1–20.

Lebour, Marie V. "The Food of Plankton Organisms." *Journal of the Marine Biological Association of the United Kingdom* 12 (1922): 644–77.

————. "The Food of Plankton Organisms. II." *Journal of the Marine Biological Association of the United Kingdom* 13, no. 1 (1923): 70–92.

Leclercq, Nicolas, Jean-Pierre Gattuso, and Jean Jaubert. "Primary Production, Respiration, and Calcification of a Coral Reef Mesocosm under Increased CO_2 Partial Pressure." *Limnology and Oceanography* 47, no. 2 (2002): 558–64.

Leu, Ming-Yih, Pei-Jie Meng, Chao-Sheng Huang, Kwee Siong Tew, Jimmy Kuo, and Chyng-Hwa Liou. "Spawning Behaviour, Early Development and First Feeding of the Bluestriped Angelfish [*Chaetodontoplus septentrionalis* (Temminck & Schlegel, 1844)] in Captivity." *Aquaculture Research* 41, no. 9 (2010): e39–e52.

Lévi-Strauss, Claude. *The Savage Mind*. Chicago: University of Chicago Press, 1966.

"Live Rock Hitchhikers." ARC Reef Marine Research Laboratory, March 2, 2020. https://arcreef.com/live-rock-hitchhikers/.

Louson, Eleanor. "Never before Seen: Spectacle, Staging, and Story in Wildlife Film's Blue-Chip Renaissance." PhD diss., York University (ON), 2018.

Lucier, Paul. "Court and Controversy: Patenting Science in the Nineteenth Century." *British Journal for the History of Science* 29, no. 2 (1996): 139–54.

Luckett, Christopher, Walter H. Adey, Janice Morrissey, and Donald M. Spoon. "Coral Reef Mesocosms and Microcosms: Successes, Problems, and the Future of Laboratory Models." *Ecological Engineering* 6, nos. 1–3 (1996): 57–72.

Lynch, Michael. "The Production of Scientific Images: Vision and Re-vision in the

History, Philosophy, and Sociology of Science." In *Visual Cultures of Science: Rethinking Representational Practices in Knowledge Building and Science Communication*, edited by Luc Pauwels, 26–40. Hanover, NH: Dartmouth College Press, 2006.

MacGregor, Arthur. *Naturalists in the Field: Collecting, Recording and Preserving the Natural World from the Fifteenth to the Twenty-First Century*. Leiden: Brill, 2018.

Machado, Pedro, Steve Mullins, and Joseph Christensen, eds. *Pearls, People, and Power: Pearling and Indian Ocean Worlds*. Athens: Ohio University Press, 2020.

Madhu, K., and Rema Madhu. "Captive Spawning and Embryonic Development of Marine Ornamental Purple Firefish *Nemateleotris decora* (Randall & Allen, 1973)." *Aquaculture* 424 (2014): 1–9.

Malindine, Jonathan. "Prehistoric Aquaculture: Origins, Implications, and an Argument for Inclusion." *Culture, Agriculture, Food and Environment* 41, no. 1 (2019): 66–70.

Marhaver, Kristen L., Mark J. A. Vermeij, and Mónica M. Medina. "Reproductive Natural History and Successful Juvenile Propagation of the Threatened Caribbean Pillar Coral *Dendrogyra cylindrus*." *BMC Ecology* 15, no. 1 (2015): 1–12.

Martell, L., S. Piraino, C. Gravili, and F. Boero. "Life Cycle, Morphology and Medusa Ontogenesis of *Turritopsis dohrnii* (Cnidaria: Hydrozoa)." *Italian Journal of Zoology* 83, no. 3 (2016): 390–99.

Martínez, Alejandro. "'A Souvenir of Undersea Landscapes': Underwater Photography and the Limits of Photographic Visibility, 1890–1910." *História, Ciências, Saúde-Manguinhos* 21 (2014): 1029–47.

Martin-Smith, Keith M., and Amanda C. J. Vincent. "Seahorse Declines in the Derwent Estuary, Tasmania in the Absence of Fishing Pressure." *Biological Conservation* 123, no. 4 (2005): 533–45.

Matsuda, Hirokazu, and Taisuke Takenouchi. "Development of Technology for Larval Culture in Japan: A Review." *Bulletin of Fisheries Research Agency* 20 (2007): 77–84.

Matsuda, Matt K. *Pacific Worlds: A History of Seas, Peoples, and Cultures*. Cambridge: Cambridge University Press, 2012.

Maynard, Jeffrey, Ruben Van Hooidonk, C. Mark Eakin, Marjetta Puotinen, Melissa Garren, Gareth Williams, Scott F. Heron, et al. "Projections of Climate Conditions That Increase Coral Disease Susceptibility and Pathogen Abundance and Virulence." *Nature Climate Change* 5, no. 7 (2015): 688–94.

McCray, W. Patrick. "Amateur Scientists, the International Geophysical Year, and the Ambitions of Fred Whipple." *Isis* 97, no. 4 (2006): 634–58.

McGrogan, Douglas G. "Mirror-Box for Photographing Small Fishes." *Copeia*, no. 4 (1990): 1174–76.

McJones, Crabs. "What Is That!! A R2R Guide to Common New Tank Hitchhikers." Reef2Reef, Hitchhiker and Critter ID. https://www.reef2reef.com

/threads/what-is-that-a-r2r-guide-to-common-new-tank-hitchhikers .443382/.

McLain, Rebecca J., Harriet H. Christensen, and Margaret A. Shannon. "When Amateurs Are the Experts: Amateur Mycologists and Wild Mushroom Politics in the Pacific Northwest, USA." *Society & Natural Resources* 11, no. 6 (1998): 615–26.

McMillan, N. F., and W. J. Rees "Maude Jane Delap." *Irish Naturalists' Journal* 12, no. 9 (January 1958): 221–22.

Meyers, Jason R. "Zebrafish: Development of a Vertebrate Model Organism." *Current Protocols Essential Laboratory Techniques* 16, no. 1 (2018): e19.

Milne, Lorus J. "A Simple, Thin Aquarium." *Science* 93, no. 2418 (1941): 432.

Minteer, Ben A., Jane Maienschein, and James P. Collins, eds. *The Ark and Beyond: The Evolution of Aquarium and Zoo Conservation*. Chicago: University of Chicago Press, 2018.

Mitman, Gregg. "Cinematic Nature: Hollywood Technology, Popular Culture, and the American Museum of Natural History." *Isis* 84, no. 4 (1993): 637–61.

Moe, Martin A. "Breeding the Clownfish, *Amphiprion ocellaris*." *Salt Water Aquarium* 9, no. 2 (March/April, 1973): 1–32.

———. *Breeding the Orchid Dottyback,* Pseudochromis fridmani*: An Aquarist's Journal*. Plantation, FL: Green Turtle, 1997.

———. "A Message from Martin Moe." Rising Tide Conservation (blog), December 1, 2011. https://www.risingtideconservation.org/a-message-from -martin-moe/.

———. "The Way We Were: 1973: Breeding the Clownfish, *Amphiprion ocellaris*." *Advanced Aquarist* 11 (February 8, 2012). https://www.advancedaquarist.com /2012/2/breeder.

Moe, Martin A., Robert M. Ingle, and Raymond H. Lewis. *Pompano Mariculture: Preliminary Data and Basic Considerations*. St. Petersburg: Florida Board of Conservation Marine Laboratory, 1968.

"Moon Jellyfish Blog: Tips and Tricks to Keeping Jellyfish as Pets." Sunset Marine Labs. https://moonjellyfishblog.com.

Moorhead, Jonathan A., and Chaoshu Zeng. "Development of Captive Breeding Techniques for Marine Ornamental Fish: A Review." *Reviews in Fisheries Science* 18, no. 4 (2010): 315–43.

Moreau, Marie-Annick, H. J. Hall, and A. C. J. Vincent. *Proceedings of the First International Workshop on the Management and Culture of Marine Species Used in Traditional Medicines, July 4–9, 1998, Cebu City, Philippines*. Montreal: Project Seahorse, McGill University, 2000.

Muka, Samantha K. "Conservation Constellations: Aquariums in Aquatic Conservation Networks." In *The Ark and Beyond: The Evolution of Aquarium and Zoo Conservation*, edited by Ben A. Minteer, Jane Maienschein, and James P. Collins, 90–104. Chicago: University of Chicago Press, 2018.

———. "The Evolution of a Reef Aquarium." Smithsonian Institution, Ocean,

October 2017. https://ocean.si.edu/ecosystems/coral-reefs/evolution-reef-aquarium.

———. "Historiography of Marine Biology." In *Handbook of the Historiography of Biology*, edited by Michael R. Dietrich, Mark E. Borrello, and Oren Harman, 435–59. Cham, Switzerland: Springer, 2021.

———. "Illuminating Animal Behavior: The Impact of Malleable Marine Stations on Tropism Research." in *From the Beach to the Bench: Why Marine Biological Studies?*, edited by Jane Maienschein, Karl Matlin, and Rachel Ankeny, 119–43. Chicago: University of Chicago Press, 2020.

———. "Imagining the Sea: The Impact of Marine Field Work on Scientific Portraiture." In *Soundings and Crossings: Doing Science at Sea, 1800–1970*, edited by Katharine Anderson and Helen Rozwadowski, 247–76. Sagamore Beach, MA: Science History, 2016.

———. "The Right Tool and the Right Place for the Job: The Importance of the Field in Experimental Neurophysiology, 1880–1945." *History and Philosophy of the Life Sciences* 38, no. 3 (2016): 1–28.

———. "Taking Hobbyists Seriously." *Journal for the History and Philosophy of Biology* (forthcoming).

———. "Trashing the Tanks." *American Scientist* 106, no. 6 (2018): 340–44.

———. "Working at Water's Edge: Life Sciences at American Marine Stations, 1880–1930." PhD diss., University of Pennsylvania, 2014.

Munns, David P. D. *Engineering the Environment: Phytotrons and the Quest for Climate Control in the Cold War*. Pittsburgh, PA: University of Pittsburgh Press, 2017.

Nagabhushanam, A. K. "Feeding of a Ctenophore, *Bolinopsis infundibulum* (O. F. Müller)." *Nature* 184, no. 4689 (1959): 829–29.

Neely, Karen Lynn, and Cynthia Lewis. "Rapid Population Decline of the Pillar Coral *Dendrogyra cylindrus* along the Florida Reef Tract." *Frontiers in Marine Science* 8 (April 2021). https://doi.org/10.3389/fmars.2021.656515.

Neely, Karen Lynn, Cynthia Lewis, A. N. Chan, and I. B. Baums. "Hermaphroditic Spawning by the Gonochoric Pillar Coral *Dendrogyra cylindrus*." *Coral Reefs* 37, no. 4 (2018): 1087–92.

Nonaka, M., A. H. Baird, T. Kamiki, and H. H. Yamamoto. "Reseeding the Reefs of Okinawa with the Larvae of Captive-Bred Corals." *Coral Reefs* 22, no. 1 (2003). https://doi.org/10.1007/s00338-003-0281-x.

Nutch, Frank. "Gadgets, Gizmos, and Instruments: Science for the Tinkering." *Science, Technology, & Human Values* 21, no. 2 (1996): 214–28.

Nyhart, Lynn K. *Modern Nature: The Rise of the Biological Perspective in Germany*. Chicago: University of Chicago Press, 2009.

Observer Staff. "Jellyfish Tanks, Funded 54 Times over on Kickstarter, Turn Out to Be Jellyfish Death Trap." Obersever.com. March 15, 2012. http://observer.com/2012/03/jellyfishtanksfunded54timesoveronkickstarterturnouttobe jellyfishdeathtraps/ (accessed May 17, 2017; no longer posted).

Oceans, Reefs, and Aquariums (ORA). "Grube's Gorgonian—New from ORA." ORA, October 30, 2011. https://www.orafarm.com/blog/2011/10/30/grubes-gorgonian-new-from-ora/.

Odum, Howard T., and Eugene P. Odum. "Trophic Structure and Productivity of a Windward Coral Reef Community on Eniwetok Atoll." *Ecological Monographs* 25, no. 3 (1955): 291–320.

Ogawa, Thomas, and Christopher L. Brown. "Ornamental Reef Fish Aquaculture and Collection in Hawaii." *Aquarium Sciences and Conservation* 3, no. 1–3 (2001): 151–69.

Olivotto, Ike, Scott A. Holt, Oliana Carnevali, and G. Joan Holt. "Spawning, Early Development, and First Feeding in the Lemonpeel Angelfish *Centropyge flavissimus*." *Aquaculture* 253, no. 1–4 (2006): 270–78.

Olivotto, Ike, Miquel Planas, Nuno Simões, G. Joan Holt, Matteo Alessandro Avella, and Ricardo Calado. "Advances in Breeding and Rearing Marine Ornamentals." *Journal of the World Aquaculture Society* 42, no. 2 (2011): 135–66.

Onaga, Lisa. "A Matter of Taste: Making Artificial Silkworm Food in Twentieth-Century Japan." In *Nature Remade: Engineering Life, Envisioning Worlds*, edited by Luis A. Campos, Michael R. Dietrich, Tiago Saraiva, and Christian C. Young. 115–34. Chicago: University of Chicago Press, 2021.

"On the Rate of Growth of Stony Corals." Folders 1–3, Box 6, Accession 99–124, T. Wayland Vaughan Papers, Smithsonian Institution Archives, Washington, DC.

Ormestad, Mattias, Aldine Amiel, and Eric Röttinger. "Ex-situ Macro Photography of Marine Life." In *Imaging Marine Life: Macrophotography and Microscopy Approaches for Marine Biology*, edited by Emmanuel Reynaud, 210–33. Weinheim an der Bergstrasse: Wiley, 2013.

Ottinger, Gwen. "Buckets of Resistance: Standards and the Effectiveness of Citizen Science." *Science, Technology, & Human Values* 35, no. 2 (2010): 244–70.

Packer, Kathryn, and Andrew Webster. "Patenting Culture in Science: Reinventing the Scientific Wheel of Credibility." *Science, Technology, & Human Values* 21, no. 4 (1996): 427–53.

Page, Charles Nash. *Aquaria: A Treatise on the Food, Breeding, and Care of Fancy Goldfish, Paradise Fish, Etc.* Des Moines, IA: published by the author, 1898.

Pawley, Emily." Feeding Desire: Generative Environments, Meat Markets, and the Management of Sheep Intercourse in Great Britain, 1700–1750." *Osiris* 33, no. 1 (2018): 47–62.

———. "The Point of Perfection: Cattle Portraiture, Bloodlines, and the Meaning of Breeding, 1760–1860." *Journal of the Early Republic* 36, no. 1 (2016): 37–72.

Pham, Nancy Kim, and Junda Lin. "The Effects of Different Feed Enrichments on Survivorship and Growth of Early Juvenile Longsnout Seahorse, *Hippocampus reidi*." *Journal of the World Aquaculture Society* 44, no. 3 (2013): 435–46.

Pickstone, John V. *Ways of Knowing: A New History of Science, Technology, and Medicine*. Chicago: University of Chicago Press, 2001.

Pilnick, Aaron R., Keri L. O'Neil, Martin Moe, and Joshua T. Patterson. "A Novel System for Intensive *Diadema antillarum* Propagation as a Step towards Population Enhancement." *Scientific Reports* 11, no. 1 (2021): 1–13.

Pletcher, T. F. "A Portable Aquarium for Use at Sea to Photograph Fish and Aquatic Life." *Journal of the Fisheries Board of Canada* 23, no. 8 (1966): 1271–75.

Pouil, Simon, Michael F. Tlusty, Andrew L. Rhyne, and Marc Metian. "Aquaculture of Marine Ornamental Fish: Overview of the Production Trends and the Role of Academia in Research Progress." *Reviews in Aquaculture* 12, no. 2 (2020): 1217–30.

Powell, David C. *A Fascination for Fish: Adventures of an Underwater Pioneer.* Berkeley: University of California Press, 2001.

Prodger, Phillip. *Darwin's Camera: Art and Photography in the Theory of Evolution.* Oxford: Oxford University Press, 2009.

Rader, Karen. *Making Mice: Standardizing Animals for American Biomedical Research, 1900–1955*. Princeton, NJ: Princeton University Press, 2004.

Randall, John E. "A Technique for Fish Photography." *Copeia*, no. 2 (1961): 241–42.

Rasmussen, Nicolas. *Picture Control: The Electron Microscope and the Transformation of Biology in America, 1940–1960*. Stanford, CA: Stanford University Press, 1999.

Redman, Samuel J. *Bone Rooms: From Scientific Racism to Human Prehistory in Museums*. Cambridge, MA: Harvard University Press, 2016.

Rees, W. J., and F. S. Russell. "On Rearing the Hydroids of Certain Medusae, with an Account of the Methods Used." *Journal of the Marine Biological Association of the United Kingdom* 22, no. 1 (1937): 61–82.

Rehbock, Philip F. "The Victorian Aquarium in Ecological and Social Perspective." In *Oceanography: The Past,* edited by M. Sears and D. Merriman, 522–39. New York: Springer, 2012.

Reynaud, Stéphanie, Nicolas Leclercq, Samantha Romaine-Lioud, Christine Ferrier-Pagés, Jean Jaubert, and Jean-Pierre Gattuso. "Interacting Effects of CO_2 Partial Pressure and Temperature on Photosynthesis and Calcification in a Scleractinian Coral." *Global Change Biology* 9, no. 11 (2003): 1660–68.

Rinkevich, Baruch. "Conservation of Coral Reefs through Active Restoration Measures: Recent Approaches and Last Decade Progress." *Environmental Science & Technology* 39, no. 12 (2005): 4333–42.

Rinkevich, Baruch. "Restoration Strategies for Coral Reefs Damaged by Recreational Activities: The Use of Sexual and Asexual Recruits." *Restoration Ecology* 3, no. 4 (1995): 241–51.

Rinne, John N., and Martin D. Jakle. "The Photarium: A Device for Taking Natural Photographs of Live Fish." *Progressive Fish-Culturist* 43, no. 4 (1981): 201–4.

Ritvo, Harriet. *The Animal Estate: The English and Other Creatures in the Victorian Age*. Cambridge, MA: Harvard University Press, 1987.

———. "Pride and Pedigree: The Evolution of the Victorian Dog Fancy." *Victorian Studies* 29, no. 2 (1986): 227–53.

———. "Race, Breed, and Myths of Origin: Chillingham Cattle as Ancient Britons." *Representations* 39 (1992): 1–22

Robison, Bruce H., Kim R. Reisenbichler, James C. Hunt, and Steven H. D. Haddock. "Light Production by the Arm Tips of the Deep-Sea Cephalopod *Vampyroteuthis infernalis*." *Biological Bulletin* 205, no. 2 (2003): 102–9.

Roos, Anna Marie Eleanor. *Goldfish*. London: Reaktion Books, 2019.

Rose, Hilary. "Hand, Brain, and Heart: A Feminist Epistemology for the Natural Sciences." *Signs* 9, no. 1 (1983): 73–90.

Rosen, Brian Roy. "Darwin, Coral Reefs, and Global Geology." *BioScience* 32, no. 6 (1982): 519–25.

Rosenberg, Gabriel N. "No Scrubs: Livestock Breeding, Eugenics, and the State in the Early Twentieth-Century United States." *Journal of American History* 107, no. 2 (2020): 362–87.

Ross, Richard. "Everyone Can Do Science." Skeptical Reefkeeping, *Reefs Magazine*, June 2, 2016. http://packedhead.net/2016/skeptical-reefkeeping-xiv-everyone-can-do-science/.

Rothfels, Nigel. "(Re)introducing the Przewalski's Horse." In *The Ark and Beyond: The Evolution of Zoo and Aquarium Conservation*, edited by Ben Minteer, Jane Maienschein, and James P. Collins, 77–89. Chicago: University of Chicago Press, 2018.

Rozwadowski, Helen M. "Playing by—and on and under—the Sea: The Importance of Play for Knowing the Ocean." In *Knowing Global Environments: New Historical Perspectives on the Field Sciences*, edited by Jeremy Vetter, 162–89. New Brunswick, NJ: Rutgers University Press, 2010.

———. *Vast Expanses: A History of the Oceans*. London: Reaktion Books, 2018.

Rubin, Gayle. "The Traffic in Women: Notes on the 'Political Economy' of Sex." *Toward an Anthropology of Women*, edited by Rayna Rapp, 157–210. New York: Monthly Review Press, 1975.

Ruiz-Ramos, Dannise V., Edwin A. Hernández-Delgado, and Nikolaos V. Schizas. "Population Status of the Long-Spined Urchin *Diadema antillarum* in Puerto Rico 20 Years after a Mass Mortality Event." *Bulletin of Marine Science* 87, no. 1 (2011): 113–27.

Russell, F. S. "Dr. Marie V. Lebour." *Journal of the Marine Biological Association of the United Kingdom* 52, no. 3 (August 1972): 777–88.

Sanders, Clinton R. "Working Out Back: The Veterinary Technician and 'Dirty Work.'" *Journal of Contemporary Ethnography* 39, no. 3 (2010): 243–72.

Schmidt, Susanne K., Raymund Werle, K. Susanne, and Trevor Pinch. *Coordinating Technology: Studies in the International Standardization of Telecommunications*. Cambridge, MA: MIT Press, 1998.

Schubert, Patrick, and Thomas Wilke. "Coral Microcosms: Challenges and Opportunities for Global Change Biology." In *Corals in a Changing World*, edited by Carmenza Duque and Edisson Tello Camacho, 143–75. N.p.: IntechOpen, 2017.

Scott, Joan W. "Gender: A Useful Category of Historical Analysis." *American Historical Review* 91, no. 5 (1986): 1053–75.

Shell, Hannah Rose. "Things under Water: Etienne-Jules Marey's Aquarium Laboratory and Cinema's Assembly." In *Making Things Public: Atmospheres of Democracy*, edited by Bruno Latour and Peter Weibel, 326–32. Cambridge, MA: MIT Press, 2005.

Shimura, Kazuko, Tanimura Shunsuke, and Shimazu Tsuneo. "Breeding of *Dactylometra pacifica* in Enoshima Aquarium." *Journal of Japanese Zoos and Aquariums* 30, no. 3 (1988): 76–79.

Shor, Elizabeth N. "The Role of T. Wayland Vaughan in American Oceanography." In *Oceanography: The Past*, edited by M. Sears and D. Merriman, 127–37. New York: Springer, 2012.

Shufeldt, R. W. "Experiments in Photography of Live Fishes." *Bulletin of the United States Fish Commission, no. 424* (1899): 1–5.

Smith, Hugh McCormick. *Japanese Goldfish, Their Varieties and Cultivation: A Practical Guide to the Japanese Methods of Goldfish Culture for Amateurs and Professionals*. Washington, DC: W. F. Roberts, 1909.

Smith, Jonathan. "Eden under Water: The Visual Natural Theology of Philip Gosse's Aquarium Books." Paper presented at the conference on "Nineteenth-Century Religion and the Fragmentation of Culture in Europe and America," Lancaster, England, July 1997. https://www-personal.umd.umich.edu/~jonsmith/PHGINCS.html.

Smith, Tim D. *Scaling Fisheries: The Science of Measuring the Effects of Fishing, 1855–1955*. Cambridge: Cambridge University Press, 1994.

Smith, Walton. "An Apparatus for Rearing Marine Organisms in the Laboratory." *Nature* 136, no. 3435 (August 31, 1935): 345–46.

Smothers, Ronald. "Violin Collector Known for Sale to Orchestra Sentenced to 18 Months for Tax Fraud." *New York Times*, March 22, 2005.

Sommer, F. A. "Advances in Culture and Display of *Aurelia aurita*, the Moon Jelly." In *AAZPA Regional Conference Proceedings*, Western Regional Conference, 391–96. N.p.: American Association of Zoological Parks and Aquariums, 1992.

———. "Husbandry Aspects of a Jellyfish Exhibit at the Monterey Bay Aquarium." *American Association of Zoological Parks and Aquariums Annual Conference Proceedings* (1992): 362–69.

———. "Jellyfish and Beyond: Husbandry of Gelatinous Zooplankton at the Monterey Bay Aquarium." In *Proceedings of the Third International Aquarium Congress*, edited by Chris Barrett, 249–61. Boston: New England Aquarium, 1993.

Spangenberg, Dorothy Breslin. "Cultivation of the Life Stages of *Aurelia aurita* under Controlled Conditions." *Journal of Experimental Zoology Part A: Ecological Genetics and Physiology* 159, no. 3 (1965): 303–18.

———. "Iodine Induction of Metamorphosis in *Aurelia*." *Journal of Experimental Zoology Part A: Ecological Genetics and Physiology* 165, no. 3 (1967): 441–49.

———. "A Study of Strobilation in *Aurelia aurita* under Controlled Conditions." *Journal of Experimental Zoology Part A: Ecological Genetics and Physiology* 160, no. 1 (1965): 1–9.

Sponsel, Alistair. *Darwin's Evolving Identity: Adventure, Ambition, and the Sin of Speculation*. Chicago: University of Chicago Press, 2018.

Spotte, Stephen. *Zoos in Postmodernism: Signs and Simulation*. Madison, NJ: Fairleigh Dickinson University Press, 2006.

Sprung, Julian. *Reef Notes 3:1993/1994*. Revisited and revised ed. Coconut Grove, FL: Ricordea, 1996.

Star, Susan Leigh, and James R. Griesemer. "Institutional Ecology, Translations, and Boundary Objects: Amateurs and Professionals in Berkeley's Museum of Vertebrate Zoology, 1907–39." *Social Studies of Science* 19, no. 3 (1989): 387–420.

Steinke, Dirk, Robert Hanner, and Paul D. Hebert. "Rapid High-Quality Imaging of Fishes Using a Flat-Bed Scanner." *Ichthyological Research* 56, no. 2 (2009): 210–11.

Stephens, Lester D., and Dale R. Calder. *Seafaring Scientist: Alfred Goldsborough Mayor, Pioneer in Marine Biology*. Columbia: University of South Carolina Press, 2006.

Stoddart, David R. "Darwin, Lyell, and the Geological Significance of Coral Reefs." *British Journal for the History of Science* 9, no. 2 (1976): 199–218.

Stone, Livingstone. "The Quinnat Salmon or California Salmon—*Oncorhynchus chouicha*." In *The Fisheries and Fishery Industries of the United States. Section I. Natural History of Useful Aquatic Animals*, prepared by George Brown Goode, 479–85. Washington, DC: US Government Printing Office, 1884.

Stone, Livingston. "Trout Culture." *Transactions of the American Fisheries Society* 1, no. 1 (1872): 46–56.

Swanberg, N. "The Feeding Behavior of *Beroe ovata*." *Marine Biology* 24, no. 1 (1974): 69–76.

Taylor, Joseph E., III. *Making Salmon: An Environmental History of the Northwest Fisheries Crisis*. Seattle: University of Washington Press, 2009.

Tew, Kwee Siong, Yun-Chen Chang, Pei-Jie Meng, Ming-Yih Leu, and David C. Glover. "Towards Sustainable Exhibits: Application of an Inorganic Fertilization Method in Coral Reef Fish Larviculture in an Aquarium." *Aquaculture Research* 47, no. 9 (2016): 2748–56.

Thompson, Victor D., William H. Marquardt, Michael Savarese, Karen J. Walker, Lee A. Newsom, Isabelle Lulewicz, Nathan R. Lawres, Amanda D. Roberts Thompson, Allan R. Bacon, and Christoph A. Walser. "Ancient Engineering of Fish Capture and Storage in Southwest Florida." *Proceedings of the National Academy of Sciences* 117, no. 15 (2020): 8374–81.

Townsend, Ditch. "Sustainability, Equity and Welfare: A Review of the Tropical

Marine Ornamental Fish Trade." *SPC Live Reef Fish Information Bulletin* 20 (December 2011): 2–12.

Traweek, Sharon. *Beamtimes and Lifetimes.* Cambridge, MA: Harvard University Press, 2009.

Tucker, Jennifer. *Nature Exposed: Photography as Eyewitness in Victorian Science.* Baltimore: Johns Hopkins University Press, 2005.

Tunnicliffe, Sue Dale, and Annette Scheersoi. Introduction to *Natural History Dioramas: History, Construction and Educational Role*, 1–4. Dordrecht: Springer, 2015.

US Department of Commerce, National Oceanic and Atmospheric Administration. "How Much of the Ocean Have We Explored?" National Ocean Service, NOAA, January 1, 2009. https://oceanservice.noaa.gov/facts/exploration .html.

Ushijima, Blake, Julie L. Meyer, Sharon Thompson, Kelly Pitts, Michael F. Marusich, Jessica Tittl, Elizabeth Weatherup, et al. "Disease Diagnostics and Potential Coinfections by *Vibrio coralliilyticus* during an Ongoing Coral Disease Outbreak in Florida." *Frontiers in Microbiology* (October 26, 2020). https://www.frontiersin.org/articles/10.3389/fmicb.2020.569354/full.

Van Oppen, Madeleine J. H., James K. Oliver, Hollie M. Putnam, and Ruth D. Gates. "Building Coral Reef Resilience through Assisted Evolution." *Proceedings of the National Academy of Sciences* 112, no. 8 (2015): 2307–13.

Vaughan, Thomas Wayland. "The Geologic Significance of the Growth-Rate of the Floridian and Bahaman Shoal-Water Corals." *Journal of the Washington Academy of Sciences* 5, no. 17 (1915): 591–600.

———. "The Results of Investigations of the Ecology of the Floridian and Bahaman Shoal-Water Corals." *Proceedings of the National Academy of Sciences* 2, no. 2 (1916): 95–100.

Vennen, Mareike. *Das Aquarium: Praktiken, Techniken und Medien der Wissensproduktion (1840–1910).* Göttingen: Wallstein, 2018.

Vermeulen, Niki. *Supersizing Science: On Building Large-scale Research Projects in Biology.* Boca Raton, FL: Dissertation.com, 2010.

Wessely, Christina, and Nathan Stobaugh. "Watery Milieus: Marine Biology, Aquariums, and the Limits of Ecological Knowledge circa 1900." *Grey Room* 75 (2019): 36–59.

Widmer, Chad L. *How to Keep Jellyfish in Aquariums: An Introductory Guide for Maintaining Healthy Jellies.* Tucson, AZ: Wheatmark, 2008.

Wilder, Kelley E. "Photography and the Art of Science." *Visual Studies* 24, no. 2 (2009): 163–68.

Wilkens, Peter. "An Experimental Marine Aquarium." *Marine Aquarist* 6, no. 5 (1975): 49–55.

———. "Mini-Reef." *Marine Aquarist* 7, no. 5 (1976): 37–42.

———. *The Saltwater Aquarium for Tropical Marine Invertebrates* Berlin: Engelbert Pfriem, 1973.

Williamson, Bess. *Accessible America: A History of Disability and Design*. 2 vols. New York: New York University Press, 2019.

———. "Electric Moms and Quad Drivers: People with Disabilities Buying, Making and Using Technology in Postwar America." In *Disability, Space, Architecture: A Reader*, edited by Jos Boys, 198–306. London: Routledge, 2017.

Wilson, Douglas P. "The Plunger-Jar." *Aquarist and Pondkeeper* 1 (1937): 138–39.

Wood, Elizabeth. "Collection of Coral Reef Fish for Aquaria: Global Trade, Conservation Issues and Management Strategies." Marine Conservation Society, UK, January 2001.

"World's First Coral Biobank in Port Douglas Key to Reef Survival." *Newsport*, December 17, 2019. https://www.newsport.com.au/2019/december/worlds-first-coral-biobank-in-port-douglas-key-to-reef-survival/.

Wylie, Caitlin Donahue. "'The Artist's Piece Is Already in the Stone': Constructing Creativity in Paleontology Laboratories." *Social Studies of Science* 45, no. 1 (2014): 31–55.

Xiphophorus Genetic Stock Center. Texas State University. https://www.xiphophorus.txstate.edu/about/introduction.html (accessed August 21, 2020; no longer posted).

Yan, Gregg. "Saving Nemo – Reducing Mortality Rates of Wild-Caught Ornamental Fish." *SPC Live Reef Fish Information Bulletin* 21 (June 2016): 3–7.

Yonge, Charles Maurice. *Great Barrier Reef Expedition, 1928–1929*. London: British Museum, 1930.

———. *A Year on the Great Barrier Reef: The Story of Corals and of the Greatest of Their Creations*. New York: Putnam, 1930.

Yoshihisa, Abe, and M. Hisada. "On a New Rearing Method of Common Jellyfish, *Aurelia aurita*." *Bulletin of the Marine Biological Station of Asamushi, Tohoku University* 13, no. 3 (1969): 205–9.

Zihlman, Adrienne. "The Paleolithic Glass Ceiling: Women in Human Evolution." In *Women in Human Evolution*, edited by Lori D. Hager, 105–28. London: Routledge, 1997.

Index